Beyond the Sea

Memoirs of an Irish Radio Officer

by

Jack Lynch

ORIGINAL WRITING

© 2010 Jack Lynch

All rights reserved. No part of this publication may be reproduced in any form or by any means—graphic, electronic or mechanical, including photocopying, recording, taping or information storage and retrieval systems—without the prior written permission of the author.

978-1-907179-79-2

A CIP catalogue for this book is available from the National Library.

Published by ORIGINAL WRITING LTD., Dublin, 2010.

Printed in Great Britain by the MPG Books Group,
Bodmin and King's Lynn

This book is dedicated to my late darling wife, Molly and to our wonderful and loving family.

Preface

I am sitting here in front of the computer, contemplating the format and layout for this story. Many thoughts are running through my head, and as I have never written anything like this before, my thoughts are that I will make a complete mess of it. However, I owe it to my family, grandchildren, and descendents to tell my story so that they may know something about me, what I accomplished, in short, what I got up to during most of the last eighty years.

Since I have accumulated a lot of genealogical family data over the years, I feel I need to pass this down through the generations, and hope that it will generate interest in who the Lynch's, Regans, Prendervilles, Mansfields, and Heffernans are and were, and how they fit into my version of past family events. For this reason I give them this book as a present from me to all of them.

It was very difficult trying to accumulate information and sift the facts from hearsay. This was the main reason that I decided to record all the information I know and researched. Hopefully, it will be easier for anybody following on from this point, to build on and expand the family history from these beginnings.

One great difficulty was, and still is, trying to positively establish my father's parenthood. Despite having a marriage certificate for him I could not find a birth certificate or details of his birth and suspect that he was adopted and reared by Ellen Prenderville and Michael Lynch, my grandparents. I did find details about a baby boy, John, born to a Mary Mansfield, on my daddy's birthday in the records at the home for unmarried mothers, in Midleton, Co. Cork. I believe this woman may be his blood mother. A hint that a Mansfield was involved in his parenthood led me to this conclusion, but it is not definite. It was during a conversation with my cousin, who grew up with daddy, that the name Mansfield was mentioned. However, my cousin said she was sworn to secrecy and could not divulge the details. All I could get from her was this hint about the name Mansfield, and I automatically thought that it was a man. Imagine my surprise when I saw the records at the Midleton Workhouse. However, looking closely at daddy's eyes, and those of my Nan, Ellen Prenderville, I feel there is also a Prenderville link. I wonder if one of the male members of the Prenderville family was the father, causing Ellen to adopt this baby. I'm still trying to find the truth. It's out there somewhere. Another reason that I believe daddy was adopted was I that I have since discovered that Ellen Prenderville and Michael Lynch were married in 1898 whereas daddy was born 4 years earlier.

Of course, I cannot omit the Kiely and Corcoran names, as they are the family names of my late wife, Molly and half my children's heritage. Thank God my memory is still good.

My lovely daughters occasionally accuse me of suffering from selective hearing and convenient memory loss. Every now and again I forget dates, or events. To be honest, when I am watching sport, particularly when it relates to the exploits of the Irish soccer, and rugby teams, I do tend to switch off from all other activities, and conversations. For this I make no apologies. Very little is more important than that little pastime which I think I have earned after many years of toil, sweat, labour, and worry. I think I have a 'gumboil,' or is it just my tongue in my cheek ….

Just contemplating on that word 'gumboil,' makes me think. I wonder how many of our descendants know what a gumboil is, or was. The symptoms were swollen gums, particularly around the molar teeth, and when the boil broke… Yuk, painfully disgusting! I have not heard of this complaint for many years now. Does it no longer exist? Perhaps it's simply known as a mouth ulcer? Collins Gem English Dictionary does not have it listed. Maybe it just faded away due to more hygienic oral cleanliness. In our early years, we did not know what toothpaste was. If, and when, we had a toothbrush, we used salt. Some people used soot.

Casting my mind back over the past eighty years, I find it hard to believe I have lived that long. Years pass so quickly and I feel this is indicative of my life being full, and content, with many happy times. For this I thank my late darling wife Molly, whom I met when she was fourteen years and five months old. She conned me into believing she was sixteen! For years we went out together, parted, and came back together again and finally married. We had thirty-eight years of very happy marriage, with the usual occasional fiery blip. She was the best thing that happened to me, along with the births of my children.

My love and thanks go to my children, Ann, Sean, Jane, and Susan, along with my grandchildren, Amy, Olivia, Stephen, Alana, Patrick and Kevin. God bless you all. All of you are wonderful, and have been all your lives. Molly and I never had one bit of trouble or unhappiness from any one of you.

One of my regrets in life is that Molly only lived to see two of her grandchildren; Amy and Olivia, who are both young ladies now. She would have had so much pride and joy, from seeing all our lovely grandchildren.

I remember with affection my dear parents and grandparents, from both paternal and maternal sides of the family. My thoughts too are with my brother Jeremiah, and my sisters, Eileen, Mary and Patty, who are all dead, RIP. To my surviving brother and sisters, I extend loving thoughts, and hope we will all hang around for many years to come.

My memories of Cobh, where I grew up are vivid and heart-warming. I fondly remember the many wonderful neighbours and school friends I had during my childhood and teenage years. Everybody knew everybody else from the town and surrounding areas. Those were fun days, when you could stay out late at night, or wander around anywhere in town, without fear of being molested, or abducted. Parents never had to worry about their children. Everybody looked after each other's children, and gave them the occasional 'clip across the ear,' if the child stepped out of line. This was allowed in those days. Unless somebody went overboard with punishment, all the neighbours allowed each other the right to correct their child. We had a healthy respect for 'Grown ups,' and addressed them as Mr or Mrs, as the case may have been. We were all allowed to fight our own battles.

It was a great time to be around, and the sun always seemed to shine. If I was rewarded a second chance to come into this world I would not change my youth for anything. We never had much money to spend, but there was love and togetherness in the family, and amongst friends. Going through World War Two was tough, but we survived and stuck together.

Was there a better place to grow up in? I don't think so. Even though we moaned about certain things my experience is that, having travelled around most of the world, the old saying; 'There's no place like home,' is still true. The roots are deep and strong. I was never happier than when I came back to Ireland from abroad, or when I was on leave from the sea.

Changed days! Maybe things are more open now, and publicity highlights what has been covered up in the past. We did not have television, and radio was only coming into the homes when I was a young child. My first view of television was in London, England in 1948. Then, it was a twelve-inch picture, with a magnifier glass perched in front of the small screen.

I am only beginning to realise how big a job it is, trying to remember everything, and get the details on the record, but I will now try and express truthfully, what most of my life was about, what influenced me, and what made me happy. This I will try to do in the correct sequence, as far as possible. Some parts, though, I will keep to myself. After all there must be some mystery, and we all have imaginations…

For my future Lynch generations, who may read this, and associate with it, I am actually trying to visualise what you will be like, and what kind of a world you will come in to. I sincerely hope you will be successful, and happy.

Love and God bless you all.

John,
Jack,
Dad,
Dada,
Granddad,
or by whatever else you know me.

1925. John and Julia Lynch, Wedding Photo

BEYOND THE SEA

1929; Cobh; Jack Lynch

Chapter One

When I was old enough to be curious daddy told me of a run-in he had with the Black and Tans, before he was married to mammy. He and Mr. John Joseph O'Connell, a Blacksmith who lived up the 'Rope Walk', were walking up Harbour Row, and when they were close to Kidney's Strand, near the Preaching House Steps, some Black and Tan soldiers accosted them. It was during the troubles and the soldiers were drunk. They took the hats off daddy and Mr.O'Connell and threw them on the ground. After searching them they then ordered both men to pick up the hats and go. Daddy picked up his hat, and said nothing. Mr.O'Connell refused and said, "If you were gentlemen you would pick up the hat and put it where you got it." Without warning a shot rang out and Mr.O'Connell died on the spot, shot through the heart. Daddy was twenty-seven years old. He never forgot this episode. It had a profound effect on him. He tried his best to get help, as the soldiers laughed. He kept the memoriam card of Mr O'Connell on the mantelpiece at home for years after, and this was the picture that ignited my interest to find out about it.

Mammy came to work as a barmaid in the Imperial Hotel in Cobh when she was about eighteen years old, and the romance blossomed. Mammy also had a run-in with the Tans. She was eighteen then and not yet married. She was working in the bar of the hotel, where my father met her. One evening a Tan soldier ordered a drink. He was drunk and when she gave him the drink he handed mammy a live hand grenade for payment. The pin was loose in it, and the Tan told her to hold it tight. She was petrified and shaking. He laughed and went back to his table and mates. An Officer of the regular British army was present, and took the grenade from mammy and made it safe. He had the Tan removed, amid scuffles, and had him escorted back to barracks.

In 1925, my father John and my mother Julia were married in Dromtarriff R.C. church in the Diocese of Kerry, even though the church was near Kanturk, in Co. Cork, where mammy was born on the 25th November, 1904. Daddy was born in Midleton, Co. Cork, on 23rd June, 1894, and spent most of his younger days around East Ferry and Aghada, Co. Cork. He went to school in Saleen National School, Co. Cork.

    He moved to Cobh and started to work in Haulbowline, as a labourer. Soon, he became a salesman and repairman, for Singer Sewing Machines, who had a shop at Harbour Row. Not long after he married he bought a Prudential Insurance book, and worked with this Company as an agent. Years later, the Irish Assurance Co. took over and he retired on pension, as Assistant Superintendent. Whilst retired, he took a spare time job as a driver with Shannon Car hire and Travel Company.

    Two years after my parents married I was born at, at home, No. 4 Carrignafoy Terrace, on March 5th 1927. Like all good Catholic families then I was baptised the next day by our Parish Priest, Rev. Fr. William F. Browne. I was the first male born, the second child, of six girls, and three boys in our family. There were three or four still born babies also born in the family. Eileen and I were the only Lynch children who were born at No. 4 Carrignafoy Terrace. This was a terraced house, which had one window already blanked out upstairs, before daddy moved in. In those days apparently the more windows that were in a house, the more tax that was paid. You paid for natural light! The house is still like that today. I don't remember anything about the inside or the garden of this house, as our family moved to Harbour Row, when I was about two or three years old.

    My father John Lynch was known by all as Jack. I was brought up to believe that my paternal grandfather was Michael Lynch, and my paternal grandmother, was Ellen Prenderville, though as I've said there is some doubt that Ellen was in fact my father's birth mother. Both were from East Cork. She married Michael Lynch, and lived in East Ferry, Co.Cork. She then moved to Cobh, with her husband Michael, and my father. This must have been post 1911 as the Census showed them all residing in House number 4, Garranekinnefeake, Rostellan, Co. Cork.

My predominant memory of Nan Prenderville is that her left hand was missing. I was told that a pig had bitten it off at the wrist, when she was a baby in a pram. She was left with a stump. I remember her well. When she was old and ill, she was confined to bed. Every day, including school days, I ran to the top of the three-story house, where she was in bed, and I rubbed Sloan's liniment into her back. She gave me a halfpenny and acid drops, which she kept under her pillow. The acid drops were all stuck together from being under the pillow, but I didn't care about that. Her hair was white and long, and when she was well she loved sitting on a chair outside the front of the house and talking to people passing by. She wore black clothes, all the time, and had a black shawl over her shoulders. I remember daddy cried when we recited the rosary at her wake.

All my younger sisters were scared of her, because they thought she was very frightening with her long white hair and sharp features. To them she looked like a witch, but with the exception of Eileen they were much smaller and younger than me at the time. To me, she was my grandmother, and I loved going up to see her.

She died in our house on January 23rd 1938, when I was almost eleven years old. I remember her laid out in the bed, but strangely I cannot remember the funeral hearse at the house, or of it being driven to Corkbeg Cemetery in Whitegate for her burial. In fact, I cannot recall anything about the removal, or funeral, which is unusual as I was so close to her.

My sister Eileen told me, years later, that she attended the funeral, whilst mammy and I stayed at home. At the time mammy was pregnant with Jerry. Eileen also said that while Nan was dying she was hallucinating, and giving out to Eileen and mammy for not offering a cup of tea to Binna, whom she said was in the next room. Eileen tried to assure her that there was nobody in the room, and she went downstairs to tell mammy what was happening. Mammy laughed, and said to Eileen. "Wait, Jack will be in soon and will go up to her."

Binna was a young family member - a girl aged seventeen, who had died, and Nan had been heartbroken and could not forget her. I arrived home, and headed for Nan's room, only to be bawled out as well, and told to get a cup of tea for Binna. I took the stairs, two or three at a time with a loud yell, to find mammy and Eileen, doubled up, laughing, at the end of the stairs.

From the year 1930, my very first memory was probably when I was about three years old. I was in bed, in the top front bedroom, in the house at Harbour Row. I awoke crying, and had a very strange feeling in my arm. On the bedroom wall I saw what I thought was a snake. Eileen, my older sister by one year and two months, was asleep in the same bed. Daddy cuddled me, and rubbed my arm, and said that it was dead as I had been lying on it while asleep. After some rubbing the feeling returned to the arm. The snake turned out to be the shadows cast by the fire burning in the grate. Mammy and daddy were sitting in front of the fire and, even though we had four bedrooms, all four of us slept in this room until we grew up some more. Each bedroom had an open fire grate.

I also clearly remember when I went to bed at night, and only candles in candleholders lit up the stairs and bedroom, I was petrified when I saw the shadows that were cast on the ceiling and stairs. Quite often, hot candle grease dripped onto the floor and burned my hands, as I ran shaking, up or down the stairs.

Another early memory was of when I was in daddy's bed, I remember waking up to go to the toilet. I was probably about four years old, as I had not yet started school. Daddy was reading a book at the time. The next thing I remember was getting back into bed and I ducked under the blankets. I heard daddy laughing out loud. "What did you get up for?" He asked me, as he peered through his reading glasses. "I went to the lavatory," I answered, and he then burst into another peal of laughter.

"You did not go to the lavatory! You stood at the end of the bed and you tried to pull your trousers over your head, and when you couldn't do it, you came back to bed" he said. He was still laughing as I fell asleep.

I think my first memory of Christmas day was about 1933. Eileen, mammy, and I were in the sitting room and the fire was blazing. I can't recall where Anthony and Patty were, as they were only babies at the time. The crepe-paper decorations, fairy lights, and the Christmas tree were all in place. The room looked wonderful, but daddy was nowhere to be seen. Suddenly, the door opened and Santa came in, with his bag on his back. Eileen and I were mesmerised, and we didn't have any idea he was going to call on us. We joyfully jumped around his bag as he took out toys and sweets. This was great! His voice was very deep and he asked us if we were good children, of course we said "Yes."

He hugged us and, after some time, he started to leave the room to visit other children. As he went upstairs I stood and watched him. I asked mammy where he was going and mammy said, "He's going up to the chimney in the room where there's no fire burning."

As he got up about five steps, Santa's outer garment became raised, and I noticed that he was not wearing Santa boots but black shoes which looked familiar. To crown it all, he was also wearing yellow and black speckled socks, similar to socks which daddy used to wear. I shouted to mammy, "Santa is wearing daddy's socks and shoes."

Santa froze on the stairs, and mammy took a fit of laughing, which she could not control. The tears ran down her face, and I could not understand the reason at the time. Eileen too looked on in bewilderment, whilst Santa went upstairs very fast. I thought I saw him shaking as he half-ran up the stairs. The best laid plans of mice and men, had come crashing down. Even at this point in time, I did not suspect who Santa was. Eventually, daddy arrived, and I told him that Santa had worn his socks and shoes. He said, "I left them out for Santa, because Santa's feet got wet on the way to the house."

Did I detect a glance at mammy, and a wry smile? I believed but never forgot.

Like all very young children it was easy to amuse us and sometimes we did not know right from wrong. Tommy Dodd was a dwarf who regularly came along Harbour Row. We were frightened of him, but this didn't stop us teasing him, much to my disgust when I now think about it." "Holy Mary, Mother of God, pray for me and Tommy Dodd." was just one of our taunts. On one occasion, two of us teased him, and he chased us. We were very small, and I remember he gained on us with his bandy legs working overtime as he chased us to the door of my home. I was terrified, and ran to mammy and she put her arms around my friend, and me, as Tommy came in the hall. Poor mammy was as terrified as we were,

"Jesus, he'll kill us all." she kept saying.

I felt her body shaking with fear as she held us very tight. Tommy stopped, looked at us for a few seconds, and turned towards the front door. We waited anxiously as he went out of the hall. Then we all breathed again. The poor man was harmless, but was annoyed at us. He had to put up with so much of this cruel mockery. My friend and I never again called him names.

When my sister Eileen and I were very young mammy sat next to us told us the story of Rupert the Bear, which was in the Daily Express every day. For me, Rupert was a live person, as well as a bear. My eyes saw his every movement, which were not depicted in the newspapers. The words of the story made him real to me. To this day I smile whenever I see his picture in papers, or magazines. Nowadays, his check trouser is in colour, and the same soft face looks out at me. We looked forward to these stories, and it's really the only newspaper story that I can remember during my early childhood, before I eventually read 'Mutt and Jeff,' in the Evening Echo. From this I progressed in later years to reading, and swapping, the Dandy, Beano, Our Boys, and many more comic books. Of course I can't forget Pudsey. I think he was a character in the religious magazine, Far East. Other reading materials in the house were; The Cork Examiner, the Holly Bough, Ireland's Own and The Messenger.

One year before Anthony and Patricia were born in 1931, daddy was told that it would be too dangerous for mammy to have any more children, as she had a very difficult time delivering another set of the twins in 1930. In fact, she received the Last Rites of the church, and spent three months in hospital after the birth of the first set of still born twins. What happened? She had five more children, and some further still born babies. Daddy was told that she should wait five years before having any more children. In total she had thirteen children. So much for, "Increase and multiply, and fill the whole earth." On one occasion, I was in the kitchen, when a school friend of daddy's arrived, and daddy spoke quietly to him. I did not overhear what was said, but daddy left the kitchen and returned with a white shoebox. His voice was low and shaky as he spoke to his friend John Will Lyons. I heard a few words about a stillborn baby. There were tears in daddy's eyes, and he quickly put the cover on the box and returned upstairs. I was not aware that mammy had been in labour, or that she had a baby at this time, but either Dr. O'Connor, or Dr Hennessy had been in the house. When I was older, there were times when I witnessed mammy fainting, on the kitchen floor, and did not realise that she was going through the change of life.

I have memories of mammy at other times washing our clothes with bars of Sunlight soap, and using the Swedish scrubbing board to get rid of the grime and dirt. The Swedish scrubbing board was a wooden board with a series of contoured ribs, running horizontal to the tub, where the washing took place. The tub was galvanized. The clothes were rubbed up and down the ribs to dislodge any dirt, and to allow the soap to penetrate the garment or cloth. There were no rubber gloves in those days, and precious little hand creams were available, so mammy's hands were rough, and sore, from the constant contact with this hard soap and cold water. She had to boil the water in a kettle on the stove whenever she had washing to do. The clothes were rinsed out in cold water, in winter and in summer. She used the mangle out in the yard to get out as much surplus water as possible from the clothes. The mangle was a device with two wooden rollers, and a handle. The wet clothes were fed through the rollers by turning the handle. It was hard work, particularly in the winter months. The babies' nappies were made from terry towelling, and these were soaked, washed, dried, aired and rewashed, over and over again. What a soul-destroying job! In those days there were no electric gadgets, or central heating, and we used coal, turf, and wood fires to keep warm. Mammy, and other women, tried to dry the clothes with these manual and primitive aids. Remember, in those days there were large families, so there were plenty of nappies to be recycled. There were nine of us little ones in our family alone.

    This routine carried on when my brothers and sisters arrived, and while mammy was pregnant. Dr John O'Connor seemed to be in the house a lot, especially when the liners arrived from America. Liners constantly arrived in the harbour so the chances were that one or two would be anchored whilst mammy was giving birth. We were told that each new baby arrived on the liner from America. In theory this makes us all Yanks! Other babies were found under cabbage leaves, but no Lynch baby was - according to the stories we were told whilst growing up.

    Besides breast milk we were all given milk straight from the cow, with no pasteurising, homogenising, or processing of any kind. Solid food was made from bread, and sugar, mixed in milk, and this was called 'goody.' I loved this food, and continued eating it for years. I always had a sweet tooth!

JACK LYNCH

1932 Twins- Patricia and Anthony Lynch, with Mammy, Eileen and Jack

1931; Lynch family; L-R Mammy & Patricia, Jack, Daddy, Eileen, Ellen & Anthony

9

BEYOND THE SEA

1933; Jack in Killarney

1935; Jack Lynch, Snr. in Cork

Chapter Two

As children we were all very active. We used to spend hours playing outdoors, away from the house and exploring the areas around us. Although, we also engaged in more structured physical activities from time to time that children these days tend to enjoy. I remember when I was very small, that daddy used to take Eileen and me to the States Hotel, (now the Commodore Hotel) where we did various gymnastic and running exercises. I have a memory of a man called Fogy Lynch (no relation), who took children for these exercises. I think he had served as a gym instructor in the British army in the past and was now retired, or at least that's what I remember about him. We climbed ropes, which were not too high, ran up and down the length of a long room, and picked up old broken type gas fire mantles, which he used for the racing exercises. He even tried to teach us fencing! It was an unusual type of exercise class, there's no doubt about that, but we loved it!

In those days, with all the freedom we had, both because mammy had so many children to mind and because of the time it was children were left to their own devices much more than they are today and accidents were common occurrences. As a very young lad, probably when I was around five or six years old, I was playing down in the bushes, in the Bath's Quay, and I jumped up to catch the top of a small wall, at the Bath's gate. I was trying to get out of the bushes and onto the road. The top of the wall was curved, smooth, and difficult to grip especially with a small hand like mine. There was a strip of barbed wire nailed to the wall, which Roger Cooney, the caretaker, put there to stop us going into the bushes, and as I jumped up my hand slipped on the curve, and the ball of my right thumb got caught in the barbwire which ripped my thumb open. The cut was very deep, and bled profusely. Somebody heard me crying, pulled me over the wall, and took me home. Mammy was demented when she saw all the blood and ran up the street to Eddie Twomey's Post Office at No 41, and phoned for the doctor.

Dr. Hennessy came, yet again as he seemed to be in our house a great deal, and when he stopped the bleeding he said he could see my 'black pudding'. He was referring to the black vein, which was very visible. He then said he would have to stitch it. I immediately went bananas, and cried "No, No". After I calmed down, he said he would not stitch it and did a great job on closing the wound without taking a needle and thread to it. I still have the scar to this day and it's almost two inches long!

Christmas for us children was a major event, we looked forward to it from October on and it was a very special time for our family; full of tradition and ritual and lots of fun. On Christmas Eve, mammy and daddy used to start preparing for the Christmas dinner. My uncle and aunts would arrive with the goose from Kanturk. This was an annual event and we all looked forward to it. The goose was plucked, singed, and cleaned. A flaming piece of newspaper was used to singe and remove any bits of feathers or down which was not fully removed from the carcass during plucking. The giblets were used for making soup and gravy. The goose was stuffed, and the ends were sewn. I never once tasted a turkey for Christmas while I was in Cobh.

Bastable and caraway seed cakes also came with our uncles and aunts from Kanturk, though sometimes they were sent ahead by postal delivery, in which case we knew before opening the package that they would be all crumbs by the time we got them and were devoured with the same relish as if the cakes had been whole. My uncle, John Riordan from Fota, brought potatoes and other vegetables, as well as edible chestnuts when he came to us. Uncle John was a gamekeeper in Fota estate and had been badly gored by a bull some years back though he was now recovered. Aunty Lina used to arrive from Ballymore, with a brace of hens, or cocks, and sometimes a duck and these were gifted to us. They were a nice change from mutton, corned beef, bacon, fish, and stews, which we were used to eating.

The ham was boiled slowly and all the children tried to take pieces when it was removed from the pot. These were the tastiest pieces of ham, perhaps because they were illicit gains, and we had to be stopped short in this, otherwise there would be none left for Christmas day. Hams were always boiled then and never baked; in those early days we only had an open stove and no gas or electric cookers.

It was an annual ritual to watch mammy mixing the ingredients for the Christmas pudding. When the mixture was finished, it was put into a silk stocking and cooked. I always wondered if the stocking was washed beforehand, but never really minded, the pudding tasted delicious every time. One Christmas day mammy went into daddy's office to bring in the goose from where she had left it overnight she found that a rat had been chewing on the goose. 'Not to worry,' said daddy, as he cut away the damaged part and rewashed the goose. It might seem terribly unhygienic now but I can't remember any of us turning up our noses at it then, and not one of us ever got sick from the cooked goose. A local woman, named Lou Lake, used to call every Christmas for the 'Pope's Nose.' This is end of the goose where the tail feathers were fitted by nature, and we were all disgusted, as it was, in short, the ass of the goose.

On Christmas Eve the house was a hive of activity. People smoked, chatted, and cooked. Drinks were available for neighbours, and friends. Nobody in the family drank alcohol. Mammy very occasionally took a sherry, or port. Daddy had taken the 'Pledge' and had given up drinking after he got married. Amid all the decorations, and the Christmas tree, I enjoyed myself as I got money, presents, sweets, and kisses. Sweet memories are made of these times.

One Christmas, a British Army Officer, who was stationed in Spike Island, gave daddy a big box of surplus toys for us. The man was returning to the UK, and did not want to bring surplus toys with him. We had great fun as the toys were in perfect condition. There were pistols, games, hats and other goodies.

As the family got a little bit older, Christmas started with midnight mass, at the Cathedral, and this was a wonderful night. Everyone in the area walked to mass together and there was great chat along the way. The choir in the church had practiced for weeks for this and the beautiful singing made everyone's spirits soar. It was truly a family outing. In order to receive the Host, we fasted for twelve hours before receiving, and this meant fasting from noon for midnight mass. This was a precept of the Church, and was binding on anybody who made their first communion, and those who were not yet sixty years. After mass we walked home, nibbled some ham, drank cocoa and went to bed. Experience made us realise that tomorrow was going to be full of excitement. It was difficult to get to sleep.

Christmas day dawned and we anxiously opened our presents. We squealed with joy at our goodies and suddenly it was dangerous to be around us as some of the toys; like roller skates, and guns that fired corks or stickers, were in constant indoor use. There were no plastic toys and most were made from tin or wood. Christmas crackers, sweets, and tins of biscuits added to the excitement. Mammy and daddy enjoyed watching us, and we usually had a big fire in the grate where mammy and daddy sat and smoked their fags; Woodbines for mammy and Player's for daddy. We roared with excitement and ran around the house for the entire day thrilled with our new toys. In the evenings we all played games like Snakes and ladders, Ludo, and various card games such as Beg off my neighbour, and Donkey. As the family grew larger the excitement became intensified.

One Christmas evening the red candle was lighting on the kitchen table as it usually was and we had all gone upstairs, to the back bedroom overlooking the backyard where we were saying the rosary. Suddenly, daddy gave a yell, and shouted, "There's a fire in the kitchen."

We all rushed down, to find that one of the cats had got into the kitchen, and knocked the candle over, and it was beginning to burn the wooden table. Fortunately, no serious damage was done by the fire. However, over the years one leg of the table had taken on a distinct shape of its own. The cats had used it to sharpen their claws and it had one section, about five inches in length, reduced to half its original girth!

Easter Sunday also was a fun day, never as good as Christmas of course but still wonderful. We got Easter eggs and other goodies as gifts. We also each got two hen eggs or a duck egg. Later, in life, the Easter Holidays from school were the best, we got two weeks off school and it was heaven.

As a child I admired my father a great deal. He was a very strong, capable man and I looked up to him. I used to love watching him as he stropped (sharpened) his cutthroat razor on the leather strap, and wondered how he didn't cut his face more often when he shaved. Like most men, he always had one or two bits of the 'Cork Examiner,' or 'Echo' stuck to his face, where he nicked himself shaving, mind you, some of the other men in the town could have been wallpaper hangers - judging by the amount of paper stuck on their faces! Daddy used a shaving brush, and a bar of soap, to create thick creamy foam which entertained us no end. He took great pleasure in daubing our noses with the soapy brush. He waited until he knew our attention was elsewhere, and then the soap splurged our noses. No matter that he did this almost every day to us we always enjoyed it and laughed ourselves silly. It was great fun!

Despite the fun and games in the house we were still ordinary children who managed to get ourselves into trouble on the odd occasion. Eileen and I used to play together upstairs in the top bedroom. Once I tied a necktie around her neck, using it as reins to play horses. Unfortunately it was a slipknot and I hitched her to the dresser to wait whilst I went to the loo. While I was away she pulled on the tie, and naturally the noose got tighter. Suddenly, I heard mammy calling daddy and I rushed up to find out what was happening. I should have stayed away! Eileen's face was blue, and she was choking from the tight loop. After releasing my sister from strangulation Daddy let out a roar and came after me. As I scarpered for my life, heading for the stairs to make a quick getaway, he lashed out in his stockinged feet but as he tried to give me a kick in the rump he missed and instead his big toe caught the staircase with a loud thwack. He let out an even bigger yell and I knew I was in serious trouble. For what appeared to me to be ages, he hobbled around with a walking stick and a cracked toe. I drew scowls and grimaces when I'd ask him how his toe was today. I was not his favourite around this time!

Harbour Row, our play area, stretches between The Bench and the junction of Harbour Hill and East Beach. It is one of the few flat long stretches of road in the town and most of it overlooks the beautiful Harbour. Some houses were built on both sides of the road, thus obscuring this lovely view from the lower part of some houses. All houses are joined together, and are three stories high. Most had shop fronts, which were beautifully maintained. In those days it was the best street in the town to live in and I loved it. Our house was very central on the row and still has an unobstructed view of Cork Harbour, and the islands in it. There were about fifty-six houses on the Row including the Sailor's Home.

The 'Sailors' Home' was at the top of Harbour Row, near the Bench. Up around this area was known as 'Dead Man's Wall,' where people met. British sailors went there, but it was not as popular, and did not last as long as the 'Soldiers Home,' which was down on the Beach. Some of the houses have been knocked down since and some rebuilt in a very different way. At the Bench junction, four roads converge; one goes up East Hill, another goes up Harbour View, the third is Harbour Row, and the fourth is another long straight road called the 'Holy Ground,' which skirts the sea-front, and has a long nautical history. This is where Queen Victoria landed. Hence the name Queenstown was bestowed on the town.

There are three gates on the Row. One accesses Kidney's Strand. This was apparently named after a Mr. Kidney, who owned the 'Brickworks,' at Belvelly. The other two gates allowed people to get down to the Baths' Quay, on the waterfront. These gates have now been closed to the public, and the little pathways are overgrown with weeds. We used these extensively during my youth, and even up to my late teens. There was only one day in the year I believe when these gates were locked, to prove the 'Right of way' was not bestowed on the public. Now it looks as if this 'Right of way' is totally withdrawn.

The grownups that lived in our neighbourhood when we were children were fascinating to us. There were three sisters, the two Misses Iretons, lived on one side of the street, and the third sister lived on the other side of the street. Two used to walk elegantly up and down Harbour Row each day, dressed in Victorian clothes. They were very close to each other, and they had constant arguments with the third sister. They left an indelible mark in my memory. They wore long beautiful coloured, silk type dresses to their ankles, coloured straw hats, and umbrellas on their arms. They were gentle people and I remember one of them 'told off' daddy, for whacking me. He told her to mind her own bloody business! She was disgusted, and turned on her heels, but it stopped my whacking.

Further down the street, close to the Iretons lived a dwarf named Mr. Browne. He used to ride a boy's two-wheel bike and he fascinated and scared us in equal measure. Miss May O'Sullivan had a grocery shop and she always used the stub of a pencil and a well worn note book for those who had a little credit. It was a ritual to see May remove the stub from behind her right ear, lick the tip of the lead and then record the details of the purchase. Another lick of the pencil before it took its place behind May's ear ended the transaction. May was always good for the occasional empty tea chest which was used to store various household bits and pieces.

Down the road Mr. and Mrs Cull were Shopkeepers. Mr Cull was also the Town Clerk. Tom Farrell, a tailor, lived with the Culls and he used to make and alter the guards' uniforms. My sister Eileen used to get a shilling from him for delivering the uniforms to the guard's house. Mr Dockery had a house with a little railed front garden at No. 37. He never returned any balls we lost in his garden so we didn't like him at all. He, like many others, kept rooms for emigrants who were going to the States. Many of these people had to go through delousing prior to boarding the liners and they stayed at this house. I don't remember a Mrs. Dockery so I am assuming he was a widower or single man. Bob Forde from No.52 had helped deliver the Scott expedition to the Antarctic in 1910 to 1913. He was a

Sergeant in the shore party from the 'Terra Nova' expedition. Unfortunately, at the time, I did not understand the fantastic achievements of this man. He seemed so unassuming. Of course now I realise what it meant and wish I could go back and find out more about it from him.

At No. 40 was O'Kane's grocery shop. Denis O'Kane was a great boyhood friend of mine until he died of meningitis when he was very young. He and I used to bring gallon tins of hard boiled sweets to the school to be handed out at Christmas. We felt like kings as each of us had a tin, and gave them to the teachers for distribution. Mrs Finn had a sweet shop, which of course we couldn't pass by without paying a visit, even just to inhale the delicious smells. Ben and Redmond Purcell were 'Painters and Decorators' and did all the local work around the neighbourhood. Packie Purcell, another brother of theirs, worked in a chemist shop in East Beach. Packie always wore white gloves, even out walking. He had few friends and seemed to be alone a lot. Ben used to give renditions of "Off to Philadelphia in the morning" at all the local concerts. Lena May Connor, and her brother Willie, were shoe repairers and Lena May was the boss and everybody knew it. Even poor Willie who was about 30 years old, and an ex Merchant Marine Engineer, was like a schoolboy with her. When she yelled, he jumped. Mr Hynes had a sweet shop, located at the top of Harbour Row, near the Bench. This was a great favourite shop with children. The shop was the smallest in the street but Mr Hynes gave extra quantities to us and had a great selection of toffees, liquorice and soft drinks. Their son, Seanie Hynes, was a well-known and friendly boy.

There was Tommy Enright's barbershop, where we had to sit on a board placed across the chair arm rests, because we were too small to sit on the seat. I heard Tommy tell one youngster to get his mother to clean his hair before he would cut it. Lice crawled in the poor boy's hair. He went home crying with embarrassment. Nicholas O'Keefe was a shopkeeper, and wholesaler. My father used to buy sweets and chocolates from him to stock our shop. In Miss Dillon's Gift shop I used to buy stink bombs, and had great fun setting these off in enclosed areas, and then, of course, making myself scarce! The expressions, and looks on people's face were a source of side splitting laughter. Miss Dillon was old and frail and a lovely person. In her shop I also bought mouth organs, caps for guns, chalk crayons, and various other little things. In later years I bought clay pipes. These pipes were used by old ladies and men for their smoking. They were cheap to buy and easily replaced. Even though many had the stems broken the men continued to use them. I used them for other purposes. One time I put a new one in my mouth and the stem partly stuck to my lips. It took some time to get used to a new one. I used to fill the bowl with coal dust, plug it on top with wet clay, and heat the bowl over the fire. After some time gas used to come from the stem and I lit it. A lovely blue flame kept flickering until the gas was used up.

A little ditty we kids used to sing comes to mind when I think of the names of people, in Harbour Row. It reminds me now how the neighbours and people in the area were such a part of our lives and saddens me now that this is not the case in modern neighbourhoods these days. The song went;

> *"Miss Murphy played on the Piano,*
> *Sessarago played on the Banjo,*
> *Miss Dillon, she sang, till she fell with a bang,*
> *On the step, outside Mick Sheedy's door."*

As it happened these people lived quite close to each other, whilst our house was more towards the center of the street. Daddy bought our house from Rev. Titchburn, the local Protestant minister to whom, I was told it had been donated in a Will left by the previous owner, Mr. A. Olsen. Now that we were in occupation of this house daddy had the priest come and bless it, and say mass. This happened for a few years. In those days a lot of the locals were Protestants, and we could not take part in their religious services. We could not attend their funerals, or enter their churches but that was normal then. We had our own masses and churches and the two religions didn't cross paths.

There were two doors for entry into our house. One led into the main hall and the second was the entry door for the shop. We had a lovely big picture of St. Patrick on the left wall, and a large picture of the Eucharistic Congress, held in 1932 in Dublin, was hanging on the right wall leading into the main hall of the house. As in most houses a holy water container was at the front door. A glass door was halfway between the front door and the kitchen. Going past the glass door, the hall continued into the kitchen, which had red floor tiles and dark blue wood panelled the walls. There was a sash-window allowing a view into the backyard. At one end there was a cast iron stove and oven, which mammy regularly polished with black lead. A door from the kitchen led out to the concrete back yard. On the outside of the sash window there was the white ceramic sink, where we washed every morning, or whenever necessary. This sink was originally situated in the kitchen, at the sash-window, but daddy had the bright idea to put it outside, and make everybody miserable, including mammy, who had to do the washing outside. Under the stairs daddy kept coal, and firewood… and fleas.

I used to watch the coalman as he came in through the hall with the bag of coal on his back, and over his shoulder he grasped the front of the bag with his hand. As he approached the coal house he turned his back to the opened coalhouse and left the bag of coal fall into the space. This was accompanied by a lot of coal dust going all over the hall floor. More dust appeared as he shook the remainder of the coal and dust from the bag.

There were dangers everywhere in this house, and probably in all the houses in Cobh. Utilities were simple and in those days there was no talk of safety in the home and all that. We just got on and dealt with things. For instance, all the piping for water and gas were made of lead and Mammy used black lead polish for the stove. We had no idea then it was dangerous, but still, it didn't seem to affect us in any way. There was constant dampness on the walls in the house too. I heard this was due to the fact that sand and stones from the sea were used to build the houses.

The front door entering the shop led to a second door, about four feet further into the shop. Here again there was another picture hanging on the wall; it was large gilt framed oil painting of a wood, or forest, in Canada. This painting seemed to be part of the contents of the house when daddy bought it. Certainly, daddy would not have spent money buying it as an antique. He was not into collecting paintings! The shop had a back room, and there was a door which led to the stairs and also into the main hallway - just inside the glass door. This back room was later divided into two sections and was partitioned by beautiful timber, - which daddy got from dismantled ships in Haulbowline shipyard, and a bathroom was built in one-half of it. The bath was also bought in the shipyard and the new bathroom seemed like pure luxury to us but wasn't to be after all...

As had happened in the Carrignafoy house what daddy failed to realise was that this bathroom now incurred additional taxes that is until a tax bill arrived. He soon dismantled the bathroom rather than pay the extra tax. It's just as well because the water had to be boiled in a kettle, or pot, and brought to the bath. By the time one lot was in the iron bath it started to get cold, while waiting for the next lot of hot water. It was farcical.

Some years later, he countered the tax law, which related to bathrooms, when he converted the pantry on the first floor, into a shower. A shower would not incur any tax. This was so small we could not undress or dress inside it, and to get to it we had to go naked, with a sheet over us from the adjacent bedroom. To make matters worse it only had cold water. Nobody ever sang in it! We were reluctant to take a cold shower, particularly in winter and would much rather have continued with our piecemeal cleaning than be squeaky clean but cold. On reflection, I can't remember if there was anything protecting the walls or floor from getting soaked. I remember some form of a curtain protecting the door but there were certainly no tiles

The stairs were covered with linoleum and brass rods to keep it in place. On the way upstairs the first flight had a pantry, two bedrooms, and a sitting room. This latter room was the full width of the house and had two sash-windows overlooking the harbour. The toilet and one bedroom, which was my bedroom, were on the second floor. From this room I could see Harbour View, but could not see the harbour. Up another flight and there were two more bedrooms. One of these bedrooms also took up the full width of the house and this too had two sash-windows overlooking the harbour. The back bedroom had one window, which looked out onto the yard, and up to Harbour View.

In my room there was a constant smell of camphor, particularly from the wardrobe where camphor balls were used to keep moths at bay. There was a trunk in one corner of the room with American stickers on it. The room was sometimes occupied by an American couple, who were relatives of Nan, and they came over to her from time to time when she was alive. It was they who had brought the camphor with them. They always stayed in this room. Maurice was the man's name, the lady's name was Minnie and she was always called Aunt Minnie. On reflection I wonder if she was Minnie Mansfield whom I believe was in fact daddy's blood mother. Maurice liked his little tipple of 'Paddy' and used to go next door to Mackey's pub to indulge.

The trunk, covered with American stickers and labels, remained in the room for years after they went back and I can remember we played with it a fair bit. We often dared each other to get into it alone but would get really frightened, I'm not sure we trusted our siblings to let us out again! After they had gone back, mammy found a wet, crumpled American dollar note out in the backyard. The delight on her face was beautiful to see, and her excitement rose when she opened it and found it was a ten-dollar note. She immediately rinsed and ironed it, until it was like a new crisp bill. The visits of the Americans ceased after Nan died and I don't recollect hearing from them again. I would bank on it that they were Prendervilles from my grandmother's side of the family because Maurice was a Prenderville Christian name.

I loved looking from the sitting room of our house where I watched the yachts racing outside in the harbour. My interest in all things nautical was growing by the day. As well as the small yachts, there were the wonderful one design yachts; 'Cygnet' (red), 'Imp' (Dark Blue), 'Maureen' (white), 'Elsie' (Light Blue), 'Querda' (Yellow), 'Sybil' (Pink), 'Jap' and three others 'Colleen,' 'Betty' and 'Lynx' (Black.). I do not remember too much about these last three yachts, as they did not race too often but got to know the others and enjoyed watching them and cheering on my favourite to win. George Radley owned and sailed 'Querda' with his brother, Dr. Willie Radley. Kevin O'Regan and his brother Frank owned, and sailed, 'Cygnet.' There was great rivalry between these two crews every Wednesday when they sailed around the harbour. Another local, Jim Horgan, owned and sailed the 'Maureen'. When I was about eleven years old, I was out in this yacht once, with mammy, daddy and Jim. It was so exciting, as Jim gave me some pointers on how to sail and allowed me take the tiller for a while. It felt great and I fell in love with sailing and the open sea. The famous large cruiser yachts, 'Gull' and 'If' constantly raced in the harbour, and I loved the Bermuda rig of the 'If' which anchored off Whitepoint. There was never a dull moment in the harbour before the Emergency. On one occasion the gaff-rigged white cutter 'Gull' with the number 107 emblazoned on her mainsail, entered the harbour and was leading a big race when she ran aground on the Sand bank. High and dry, she waited for the tide to lift her, but to no avail, as other yachts passed her. If only she had gone around the Spit lighthouse instead of cutting across the Sand bank she would have been home and dry. What a dreadful end to her race. The year escapes me, but I think the race may have been the Fastnet race, now one of the most infamous boating races in Irish history due to the terrible tragedy that occurred in 1979.

From this room there was a beautiful view out to Roche's Point and beyond. I could see East Ferry, Aghada, Whitgate oil refinery and Fort Carlisle on one side, and the view continued and swept around to Fort Camden, Haulbowline, and Ringaskiddy on the other side with Spike Island, smack in the centre. When the liners came into the harbour, before the war, the view from the sitting room and top bedroom was breathtaking, and I saw them clearly, day and night. At night their lights were bright and it was a lovely sight as they lay at anchor off Whitegate. There was also a fine view of the Baths Quay and the red brick powerhouse, with its chimney rising up above the attached building. During the Emergency that building was used for storing coal, and turf and I worked here storing turf during my school holidays.

Tenders departed downtown from the quays to take emigrants and mail out to the liners, and bring immigrants, other passengers and mail back. Two of the tenders that I remember from those early years were s.s Saorstat and s.s Failte. The 'Green boat', named Victoria and Albert used to ferry people from Cork, down to Passage, and then onto Monkstown. From there it called into Cobh, and on to Aghada, before setting off again to Crosshaven. It then returned back to Cork.

Many Shipping Lines had offices in Cobh including the famous Cunard, White Star Line, and North German Lloyd. At midnight, on any New Year's Eve, it was terrific to see ships in the harbour, all lit up, and decorated with flags. Hooters, horns, and sirens all sounded on the stroke of midnight to herald in the New Year. Searchlights from the warships and Forts lit up the sky. I will never forget those sights. I loved the view and sat at the window looking out enthralled. The Spit lighthouse and the old coal hulk were off towards Cuskinny. Roche's Point lighthouse was out at the mouth of the harbour. the Bath's Quay could be seen. It was however very depressing looking out the window on a foggy day listening to the Fog-horn at Roche's Point lighthouse...

Chapter Three

When I was about seven years old daddy and mammy took me on an outing to Killarney with agents of the Prudential Insurance Co. This was a company outing for the Insurance personnel as a gesture of goodwill by the company. We went by bus. The bus was red, had a long bonnet, and was a real boneshaker. I don't know why Eileen didn't come, and the twins would have been too small at the time. The smell of the fumes, from the exhaust were overpowering and sickening and I remember that after a few miles, I got sick all over mammy. It was a long trip to Killarney for her as much as for me.

In Killarney that day, I had my photo taken, and I still have this picture in my album. Looking at this photo I wore long grey trousers that day, and I can distinctly call to mind the reason for getting them. Normally, I wore short trousers, but the long trousers covered the boils that I used to get quite often on my knees around that age, and this was probably because of a lack of vitamins. Many of the agents gave me money to buy sweets and treats for myself and I felt on top of the world. Thinking back over this another memory has just arisen of an incident that occurred that day. I had a beach ball, which daddy had bought for me and I liked to play with. One of the insurance men took it and started playing and punctured it when he kicked it. He paid me the money to buy a new one, but I held onto the money and didn't tell anyone. I think I would have been in trouble if my parents had known.

Besides days out and interesting trips we often, as a family, had some of our best times at home, all together. We had a gramophone record player which was brilliant and we loved to listen to music on it. Daddy played a selection of 78 rpm records, and these were mostly hymns, or John McCormack singing Irish ballads. This player was a mechanical wind up one that had a large heavy arm, and this was fitted with a changeable steel needle. The needles were sold in tins of one hundred, and were about ¾ inches long. These needles had to be changed for every record, otherwise they damaged the tracks of the records, because it was difficult to place it accurately into the grooves of the record, and a steady hand was needed. Many of the needles were rusty and quickly lost their sharp points. Later, when only the head of the arm remained, and the main arm went missing, I used to hold the head and needle, in my hand, and try and play a record. There were many more scratches on the records after this.

When on our adventures outside as children, roaming the streets and exploring we often had a companion in the form of our little King Charles dog whose name was Fluffy. The dog had a habit of going next door to Mackey's pub, where he sniffed the front of the door, lifted his leg, and piddled against the wall. This drove Mr.Mackey mad and he put pepper where the dog piddled hoping this would deter him. The dog came out to piddle again, sniffed the door and then ran sneezing into our house. I thought this very funny. I have no idea what happened to the dog, but I would not be surprised if daddy got rid of him, because of Mr. Mackey's complaints.

On the other side of our house at No 35 there was a tailor named Vincent Lotti. He was slightly built, wore thick lens glasses, and displayed a bushy moustache. He was a genius with a needle and thread. Daddy gave him his discarded suits so that he could make suits for me. This man performed miracles with daddy's old suits. The suits were ripped apart, and the material was turned inside out and bingo, I had a new suit. He was a great tailor, but was very short sighted. He sat crossed legged on the table near the window and only had a candle to throw any light on his work area. This was pre-electricity. It was something else to see him trying to thread a needle. The other thing I remember was his ritual of ironing the clothes. The iron was a large heavy implement made up of two parts. One solid lump of iron was put into the fire to heat up. When it was hot enough it was loaded into the hollow body of the actual iron. This gradually heated up the base of the holding section and this was used to iron the clothes. The handle got very hot, and was covered with a cloth to protect the hands. Most houses had these irons, but Vincent's was larger than the normal household irons.

Kate was Vincent's wife and together they had one son, also named Vincent. As I said Vincent (Snr) was short-sighted, and one day he called to Kate, and asked; "Kate, where are the blue tacks? I want to tack down the linoleum in the hall."

"They're in the white paper bag in the larder," Kate replied.

Vincent found the bag and proceeded to tack down the lino. After a little while he shouted to Kate;

"Kate these bloody tacks are no damn good, the heads keep coming off."

Kate came downstairs, looked in the bag and said, as she laughed; "Vince, you blind old bat, they're not tacks, they're cloves!"

Kate herself was a character. She was a small and chubby with a red nose and a constant smile on her face. She wore thick lens glasses too and more often than not she wore slippers, even when she came into our house, which she did on a regular basis. She also went into the snug next door, in Mackey's pub, for her 'little jar'. Often, I got the whiff of Paddy whiskey from her. The whiskey seemed to keep her going and she had a busy life to manage. One day she got very distressed because as she came in through our kitchen door a picture of the Sacred Heart, which we had hanging over the door, came down on top her and her head came through the picture. Fortunately, for her, there was no glass in the picture as it had fallen on another occasion, and the glass had never been replaced. Anyway, poor Kate got down on her knees and said;

"Lord I'm sorry. I'll go up and pay St. Anthony the money I promised him. I really forgot about it."

She had apparently got some favour from St. Anthony and promised to give a donation to St. Anthony's Bread for the poor and had completely forgotten. Poor Kate! She went up that day, and paid her dues and no doubt always paid them on time in the future.

Along with remembering these little snippets another big moment in my life was about to happen. I was about seven years old Daddy gave me a brand new two-wheeler Raleigh cycle. It was black and a bit too big for me at the time but I loved it. At that time there were very few cars around and only horses and carts were generally to be seen so this became my ultimate mode of transport. Anyway, I got on the bike and was scared of falling off but daddy held the saddle as I struggled to cycle. I was at full stretch reaching the pedal as my legs were too short for the bike and I couldn't sit on the saddle and reach the pedals at the same time. One day daddy let go of the saddle, without me noticing at first, and I was doing quite well cycling on my own until I decided I wanted to stop, and there was no hand to help me. I panicked and rode close to the Baths' wall hoping to be able to stop by slowing down against it. Of course it wasn't as easy as I might have thought and I scraped my knuckles badly and fell off. But, as the saying goes, you have to get right back on and after some more lessons I mastered the art of getting on, cycling, and getting off – without too many further injuries.

I adored that bike. I peddled up and down Harbour Row, day after day, and became quite good at it. The other boys were very envious and wanted 'a try' on the bike too. Eventually, the time came when I got too big for this bike. I had to improvise, because no more freebees were available, so as an ongoing hobby I collected old bicycles frames, and pram wheels and made makeshift bicycles myself! They mightn't have looked the best but they offered me the freedom I'd come to enjoy so much with my first bike. The only problem was that I needed a hill for self propulsion or alternatively to run with the bike and mount it while it was in motion.

During my early childhood years there was no electricity in the house. Something of course, that had no bearing on their lives. Gas and candlelight were the main sources of lighting. Prior to the arrival of electricity, gas lamps as well as candles, were used in our shop area. The fragile gas mantles, which glowed bright, fascinated me. Yes, we had paraffin oil lamps as well, and these smelled quite a bit, immediately on lighting, or after they were extinguished. We also used paraffin oil burning Primus stove to do the cooking, supplementing the coal and wood burning stove. I have a vivid memory of seeing the Gasman, lighting, and extinguishing the old public gas lamps. One lamp was located up at the Preaching House steps, and another was located outside Dockery's house, in Harbour Row. Twice a day he used a long pole, with a hook at the end to pull the chain which controlled the gas flow. Also on the pole there was a lighted wick which ignited the gas mantle. From dusk to dawn these gas lamps glowed and the gasman went around each morning to extinguish the lamps. I can't remember when the practice finished, but it was great that we had this lighting, even if it was quite dull. There was an overlap between the complete installation of Public electric lighting and the discontinued use of the Public gas light

It was 1925 when W.T Cosgrave first got involved in the idea of having cheap electricity in Ireland and, with Dr. Tom McLoughlin, decided to seek German know-how to build a power source for the country. Ard-na-Crusha, on the Shannon, was chosen and the Germans built the power station. Dr. Tom Mc Loughlin managed the project. However, due to the poor infrastructure in the country all the equipment and materials had to be shipped direct to Limerick. In 1927, on my arrival into this world, the ESB was granted the licence to manage the station and the country network. In 1929 the ESB started supplying electricity to the country. This led to the Industrial revolution which started in 1932 under Sean Lemass.

It was no easy task organising who should be the first to get electricity into their homes. Of course everyone wanted it right away but it was up others who were the lucky few to get the first service. The choice was made and work began by men digging holes for the poles and then planting these poles vertically into the holes. It was a back breaking exercise and the men had to strain all their muscles to achieve the objective. Transformers and wires were then attached to the poles and spurs taken to their final destination. Electricity meters and fuses were fitted in the respective houses which would have been inspected by engineers prior to installation of the equipment, and the switching on of the power came to Ireland, a momentous moment in our history.

After electricity arrived on the scene, in the mid thirties, almost every catholic house had an electric Sacred Heart lamp installed. These were a great source of comfort to the elderly people, and most of the elements were in the shape of a cross. The advent of electricity made it possible for children to play outside at night too, and people went for walks under street lights. Up to now, especially in the winter, it was pitch black, and only paraffin lamps and gas provided illumination in shop windows and dark streets. As children we were scared of 'the Bogeyman' and 'the Banshee'. People gathered in houses and there were some storytellers who frightened us with their ghost stories and stories about the fairy rings which were dangerous to tamper with. There were stories about Banshees being heard and somebody dying soon after. It terrified many superstitious people. Now it was different, even though there were dark patches, there were many more people about, more light, less dark corners and we felt safer.

The sound of the Banshee was one of the sounds of those days but another sound which rang in my ears was the snarling and screeching of wild cats. Out the back of our house there was a sheer cliff that reached up to Harbour View. Various bushes and wild shrubs grew on it and wild cats used to breed there. These cats often wandered to our back door and mammy fed them. She was the only person who could get close to them. They trusted her. It was terrible listening to the tomcats fighting during the mating seasons. Sometimes, it sounded like babies crying. I could hear them out on the shed fighting, and making a terrible racket. At that time, as a young boy, I did not know what mating was, though I knew the sounds of it upset me. One evening I was eating at the kitchen table when I felt something brush against my leg. I looked down and there was a big tomcat doing what came naturally with another cat. Being ignorant of the 'facts of life' I kicked the tom. There was an almighty screech and I thought the tom was going to have a go at me. His back arched, hairs stood up on his neck, and his eyes glared.

Fortunately, help was on hand and mammy shooed the cats out to the back. Though mammy fed the cats and seemed to like them they were still wild and troublesome to the rest of us. My brother Anthony had a canary for a while, until it was silly enough to get out of its cage and one of the cats had it for lunch.

One evening, while we were eating at tea-time, daddy looked out the back, and let a mighty shout out of him. The bushes and cliff were on fire. Somebody in Harbour View had set fire to the bushes. We had a good idea who did it, but it was forty years later, while I was home in Cobh on holidays, that a man came up to me and said that he used to live in Harbour View. Despite the forty years time lapse, I immediately recognised him. He admitted that he had set fire to the bushes. We had a good laugh, when I told him of the efforts which were made to control the fire. Guard Donovan had been living close by and he came down to assist my dad in controlling the fire. All they had were a bucket, a small bore hose and very little water pressure. I could have piddled higher than the water from the hose. I don't remember a fire engine coming, but people in Harbour View assisted by throwing buckets of water down on the fire. The fire badly damaged the shed, which was later demolished. The Insurance money paid for the wall, which was built across the back of the yard near the outside loo, to be reinstated. The wall was necessary more than ever now as the fire had left the cliff devoid of all weeds and bushes and it was a danger to us. At the time of the fire cats and rats scurried in all directions, amid all kinds of squeals.

Another shed was built and as the years passed this became a holding place for all kinds of wood, bicycles, and a storage space for everything that could not be used in the house. In this shed daddy later built a fish-smoking unit, for all the fish we caught in the harbour. It also became a breeding place for the wild cats that roamed the cliff. These cats in turn kept rats, and mice, at bay. The cats that frequented our backyard kept the mice at bay but the occasional mouse got into the house. I remember one day putting on a jacket, which had been hanging near the back door, and as I was adjusting it I felt something moving near the back of my neck. Daddy got up from the table, and said, "Hold tight, it's only a mouse," and hit me on the back with his hand. He reached into the back of my jacket and proudly revealed a dead mouse, which the cat took for a snack. I don't remember this bothering me at all!

Later, I kept two hens there and got lots of eggs. I was always afraid that the cats and rats would attack the hens and eggs, but strangely it never happened. There was one period when I thought that the hens had given up laying as I did not get any eggs for about a week. One day I was rummaging in the shed and, lo and behold, I found the missing eggs. The hens had changed their laying habits and now laid their eggs under some wood, which was stocked at the back of the shed. It was lovely to see the pyramid of brown eggs and I had two trips to collect all of them. They were spotlessly clean and did not need to be washed.

In those days hygiene was very limited and bathrooms were sparse. Most children had only one set of clothes and these were not washed too often. Very few children had underwear. The term 'high water mark' was commonly used. It referred to when a child's face was washed. Necks weren't washed, just the faces and the dirty mark between the washed face and unwashed neck was called a 'high water mark.'

Along with poor hygiene healthcare there was another part of our lives that was in need of serious review. Going to the dentist was a nightmare. Any ache was the signal to have the tooth out, regardless of the cause. No fillings, no sympathy, just the tooth pulled. The school dentist was a butcher and we put up with pain rather than let him get his needle, pliers and hands on us. We knew by the stories of the other children that nothing was worse than getting a tooth pulled so we lived constantly with toothaches and pain, it was preferable to the alternative. Other dreaded diseases were diphtheria, scarlet fever, and mumps, which also took many lives. People spoke in whispers when a new outbreak occurred. The children dreaded inoculations and I remember closing my eyes, yelping, and being scared when the time came for mine. It's funny how a little needle sets the nerves on fire. Applying iodine with a feather to the throat treated any sore throat. It was not pleasant but it worked.

Head lice were rampant and it was difficult to avoid getting them. We were all martyrs to the fine steel toothcomb, which had to be applied regularly to keep our heads clean. It was also used to remove scabs from the head. In fact, I remember daddy washing my head in methylated spirits to get rid of the nits and head lice. I just could not avoid them. All the children and adults suffered from the same afflictions. The fumes from the spirits nearly choked me but it did the trick. Paraffin oil was also used on the head. I made sure though that I kept my eyes closed all the time while the washing and combing was going on. In school I could see the damn lice crawling on children's hair and everybody was scratching himself, or herself, furiously.

From a health point of view we were all given regular doses of horrible castor oil, senna pods, and syrup of figs, to relieve constipation, and 'clean the bowels.' These medicines were harsh, and tore the stomach out of us, and the pains inflicted were worse than the constipation. All of it was a far cry from modern treatments but effective in their own brutal way.

Daddy, himself, never believed in following instructions on bottles or packets. If the dose should be one teaspoon then he figured that two, or even three, would be more beneficial. He swallowed a half bottle of Milk of Magnesia regularly, whilst remarking "that's great tack". Mrs. Cullen's headache powders were in great demand for headaches, colds or flu. The powder was put on our tongues and we had to swallow it with milk, or water. It was horrible and made us gag. If we were lucky, it was mixed on a spoon with jam and then this was easier to swallow. 'Parishes Food' was a tonic full of iron and that, plus 'Virol,' which was a mixture of Cod liver oil and malt, were given to us regularly. Hot poultices were frequently applied to 'ripen' boils, or to extract pus from wounds. This was a quite painful practice as the poultice, made up of bread soaked in very hot water, was kept pressed against the sore, or boil. Ear aches and tooth aches were very prevalent. Tooth decay and wax in the ears were the main reasons for these complaints.

Obviously, tuberculosis was a much feared affliction. Many people, including children, died from it and others were sent to sanatoriums in Sarsfield's Hospital, in Cork. Consumption is another name for Tuberculosis (TB.) I attended many funerals of school friends and adults who died from this scourge.

When I got a bit older I developed the common complaint of acne or blackheads. These were usually on my forehead and daddy took great pleasure squeezing them with his fingers or by using a metal watch key to get them out. Often he drew blood and soon my blackheads developed into horrible boils. I got very embarrassed by this and wished they would leave my face and go to my neck, under the hair so that they would be hidden. They went to my neck alright and also stayed on my face. I was told, "They will go when you get married." It was going to be a long wait!

No one had any money in those days and daddy tried anything to stretch any money he did earn. He worked long hours drumming up Ordinary Branch Insurance, which gave him good bonuses. A lot of this was done quietly, as he went outside his own territory, and if he was caught, it meant that there would be repercussions, because agents in those areas would object and complain to Head Office. Normally, he got his bonus in March and when he arrived home he'd have something nice for mammy and we all got presents. I remember he once came in with a posy of violets for mammy. I was very young but the delight in mammy's face at the time left a deep impression on me. He said, "For you, Julia" and he kissed her. Normally, he called her Sheila, but Julia was her Baptism Christian name. That was one of the only times I remember any hugs or kisses being exchanged between them in front of us.

Chapter Four

During my youth it was a rough time for mammy. Not that she was mistreated or anything. It's just that most women suffered the same dreary existence of hard manual work and lots of children to care for. Daddy did get her some help in the form of Diana Foley as the children started to arrive, we loved her. She was kind and really cared about us, I think Diana's arrival coincided with mammy's release from hospital, after the stillborn twins were born in 1929. The bill for the hospital was very costly and daddy was worried sick wondering where he would get money to pay this bill. He made a novena and on the last day of this prayer his insurance bonus arrived for the exact amount of the bill. The power of prayer was clearly present for him.

    Though times were hard we didn't really know it at the time. We had lots of love and lots of joyful times. We may have been short of money but I remember we often had toys to play with. Often they would be toys we had fashioned ourselves out of odds and ends but that never affected our enjoyment of them. Tops were the big craze when I was growing up. I loved what we called the 'racer' top. This was shaped like a mushroom, with a flat circular top, on a pointed spindle. To make them fancy looking we marked the top with coloured chalk. This gave a spectacular gleam as the top spun at high speed. We grew very adept at using the whip to achieve long distances and speeds during races.

    Another game we played was hoops. We scrounged bicycle wheel rims and used a piece of wood or a flat bedspring to beat them up the road. I preferred to use a flat bedspring, bent at an angle of about forty-five degrees. This, I wrapped around the outside of the wheel rim to pull it along. By doing this it eliminated the constant need to beat the wheel with wood. It did however cause a lot of grief to people when they heard metal scratching on metal which caused their teeth to water.

Boys played football using newspapers, which were wound into a tight ball and tied with twine. It was great fun and we carried on for many years even when we grew up in our teens. Any tin found was kicked into oblivion. Our boots soon wore out and the toe caps were badly scuffed. Polish! What's that?

Playing alleys, or marbles, was another common pastime. We played 'follow the leader' and the gully between the road and footpath was the play area for this game. The number of shots allowed was specified at the beginning. The first player started off and played his shot, using forefinger and thumb only, and he was followed in turn by the next players. Whoever fired the longest shot became the leader as the game progressed, until one eventually finished first. Another game was for the contestants to place marbles within a circle and then to use their expertise with their favourite marble to knock out some marbles from the circle. Each contestant had to toe a line and there were rules. Contestants tossed for start and if anyone knocked out a marble they had another try. However, the active player had to be quick to call his preference for the type of shot he wanted to take. 'Risers for a Trizzie,' meant that he could lift his hand up from the ground, so that he could let his marble enter the circle at an angle, and if there were about six marbles within the circle, it was possible for him to knock two or three out together. The other contestants may call, 'Knuckle down' before the contestant got his call away, and this forced him to take the shot with his thumb facing down the ground. This made it impossible to get any aim or power into the shot.

A fairy story we all loved back then concerned conkers and how if horsehair and a conker were put into a jar of water the hair would become an eel after three months. I can't tell you the time we all spent looking for these eels before we got wise. Homemade telephones were another source of fun, a far cry from the mobile phones of today. We got two empty tins and made a hole in the centre of the bottom of each tin, where we fitted a continuous piece of string into the two holes. A knot prevented the string from slipping through the hole. Then two of us moved apart, to the full length of the string, and one talked into his tin whilst the other listened to the sound through the second tin. Great fun and it really worked!

Like these days we dressed up as witches and ghosts on 'All Souls' night, or Halloween as we know it now. A lot of the games and activities we played at Halloween then are still enjoyed now. Within the house we played 'snap apple.' The whole family played this game. An apple was hung on a string from the frame of the door. Each person had to try and get a bite of the apple without hands touching it. Hands had to be kept behind their backs. This was difficult as the apple kept slipping around the face and knocking you. Also, like some children do these days at Halloween we would float an apple in a basin of water and try and get a bite of the apple using only our mouths, again with our hands behind our backs. Normally, the water was about nine inches deep, so we had to virtually go under the water to trap the apple. There was a lot of snorting and spluttering but it was great fun.

As the years went by collecting picture cards from cigarette packets became an obsession with me. Player's cigarettes were the tops for these cards. I remember there were about 50 cards in each set and it was great fun swapping with other children. The collections ranged from Footballers, Airplanes, Flags of the nations, birds, chickens, racehorses and flowers.

Though it might seem now that our times were hard we never knew any different and were probably happier than children today are. We just got on with things and didn't let anything bother us. The hard times did mean, though, that things in the house were never quite as nice as perhaps we would have liked. All renovations and DIY was usually done by the father of the house. There was never any question of bringing in someone to do any home improvements, things were fixed as they broke and as best they could be until the next time.

Though there was never any spare cash there was always lots of fun and laughter. On sunny days we would spend the whole day out the front, happy and relaxed and my memories are that the sun always seemed to shine back then. There were lots of children to play with, especially as my brothers and sisters started to arrive at a regular pace. The church quoted the bible; 'Increase and multiply, and fill the whole earth' and Irish Catholic families followed. Men loved this law; one the few religious laws then which got the thumbs up ….

Technology was never a big part of our lives then like it as now, as I comfortably type these words on my computer I recall a memory from early childhood when daddy arrived with a projector to show films for us. We were incredibly excited as it was such a novelty. The one and the only film that I remember was called 'Early Bird.' The projector had a very bright light and this made it very hot. The projector was also quite noisy while it played but the nothing could take from the thrill of watching these moving images. The cartoon film was black and white and very grainy. In the film I could see the worm as it came through the hole in the ground and the little bird watched it, ready to pounce for his meal. The same scenes were repeated, again and again, until the bird eventually got the worm. One day, the bulb suddenly got brighter, and brighter, until there was a bang, and out went the light. We never got a replacement.

The world around in those days was a giant playground. Enjoyment could be had from all and anything that entered our neighbourhood. We were good kids, I think, but like all kids we had a mischievous side to us! Roger Cooney was the caretaker of the Baths Quay grounds and we used to annoy the hell out of him by getting him to chase us at any opportunity, the thrill of running for your life from Mr Cooney had us in hysterics, our hearts beating loudly and gave us a surge of energy that kept us just a few steps ahead of him, luckily! Between Harbour Row and the Baths Quay there were bushes, with two pathways from entries about one hundred and fifty yards apart. Down the pathway there were six trees in a row, from entry gate to the Baths Quay. The bushes were part of the 'empire' of the Harbour Row boys and girls and they stretched from Kidney's Strand to the Bench. We played cowboys, Indians, soldiers, and various chasing games in these bushes. We knew each and every shortcut through these bushes; they were our domain, our kingdom. There were various tracks we had made whilst playing that kept us out of sight so it was difficult for anybody to catch us. However, we stuck mainly to the stretch between Kidney's strand and the gate opposite the Royal Liver Insurance Company's office, at number 38 Harbour Row.

I remember one great day that I spent flying through those bushes when I was a young lad. I kept chickens in our back yard and I also had a Bantam cock which my aunt had given me. The chickens wings were trimmed to stop them flying away but the cock did not suffer this ignominy. The cock escaped one day and he flew into the bushes in the Bath's Quay. He kept evading me as I chased him and soon enough the chase developed into an all out hunt as boys, girls, and adults who were nearby came to help. Eventually somebody caught him and we had to clip his wings too, much to his annoyance. Anyway, this and our daily running and chasing in the bushes annoyed poor Roger Cooney and he tried his best to catch us, but to no avail. To my knowledge no one was caught by Roger Cooney!

Many of our special days and celebrations as children are still around know, mostly the religious events that we observed. We always loved Shrove Tuesday, also known as 'Pancake Tuesday,' the day before Ash Wednesday. We called it 'Skelleging' day and it was the day when young school girls were fair game to be caught, tied with ropes, and have their heads put under water from public taps, available around the town. Something I'm sure would not go down well these days! I don't know what the origins of it were, or why it was a tradition but everyone in the town was aware that this was a common game on this day. Usually the boys were kept back in school by the teachers on this day to allow the girls to get home before we got loose. I think some of the girls looked forward to the game, as there were plenty of them who walked ever so slowly home that day who were caught and drenched amid screams and laughter.

Getting back to Roger, we gave him some consolation when we caught three girls one Shrove Tuesday, all sisters, whom we tied to the trees in the pathway to the Baths, giggling and shrieking, and then we went on to annoy Roger. We teased and taunted him until he chased us. As he ran after us he saw the girls tied up and freed them one by one – taking his vengeance out on them as he released them with a clobber. We stood a short way off doubled over laughing as the girls planned their revenge. Roger had a wire-haired terrier that frightened us by growling and running after us, we didn't know if he would bite but never let him get close enough to check.

    As we got older our toys often became more sophisticated. We got wooden boxes from the shopkeepers to make 'coaches' as we called them. We used a wooden box, a shaft, two axles, and four pram wheels plus a piece of rope and some nails to make karts that we'd recklessly speed down the hill in. I was the coachbuilder, as I was mechanically minded, and loved tinkering with tools. 'Sunlight' soapboxes were the best boxes for coaches and a various mixture of pram wheels would suffice if four identical wheels could not be scrounged. I had great fun with these carts, especially coming down hills, and there were plenty of these in Cobh. The reason I selected the 'Sunlight' soap boxes was that they were the most convenient size for me to sit comfortably in, and the sides did not interfere with my arms when I was steering the coach. If we ever needed help in making these coaches there was always a wealth of helpful adults on hand to assist. Diana Foley, who helped mammy with the cleaning in our house when we were growing up, was always interested in seeing what latest things we were getting up to. We all loved Diana and she was great fun; she played games with us and took care of us. She made artificial flowers with crepe paper and I thought she was a genius. Her father helped me to file down the thickness of some oversized axles to fit small wheels. He also carved aeroplanes, and wind mills from wood and gave them to me.

    Holidays then we always fantastic fun, we never went far but it still always felt like a world away when we did leave Cobh. During the late thirties and early forties our family went regularly to my Grandfather's farm, in Kanturk, for holidays.

The O'Regan family still have the farm in Paal, Kanturk. During my school days I spent many happy days at the farm during the summer holidays with my brother Anthony, my sister Eileen, and all my first cousins. I slept upstairs in the farmhouse when we were there and had to climb a ladder to get up to the beds. The pillow-cases and sheets we used were made from white flour bags, which were washed in hot water fortified with washing soda to remove the manufacturer's name from the bags, and they were then sewn together. Sometimes the pillow-cases were used to make undergarments for the women. Mattresses and pillows were stuffed with feathers and down and I sank comfortably into them. The farm fowl supplied all the stuffing materials. Sometimes, the quills of these small feathers stuck into me. All in all, it was very comfortable and warm and I always slept well there.

It was great to be part of the farm community then. Being among horses, donkeys, dogs, cows, bulls, chickens, geese, and ducks etc was wonderful, especially with the lovely continuous sunshine. I cannot remember one wet or windy day, during my annual holidays in Kanturk. I had so much fun drawing the hay, collecting eggs from the stalls and sheds, jumping and rolling in the hayloft, and often getting stung with nettles and thistles hidden in the hay. It was all part of the experience and made for memorable times.

Every morning I would wake up to the sounds of cocks crowing, hens clucking, cows mooing, and the odd bull roaring in the haggard. The donkey added to the chorus of sounds. Farm activity was always audible and visible. Ducks waddled around, and chickens spent their time pecking away at the earth. Pigs snorted and squealed. Dogs barked and granddad shouted at something or other. It was a haven of colour, noise and busyness that we all adored. There was also the lovely smell of burning turf and chopped wooden blocks.

My uncle Paddy, aunts, and grandfather had their chores to do every day and we'd go and help each of them at various stages throughout the long days. I'd watch as my aunts baked bastable cakes, and I'd always get my own piece of dough to make a biscuit for myself. The bastable dough was put into a three-legged pot, covered, and hung over the big open fire in the hearth, and then glowing embers were placed over the cover. Fresh red-hot embers were added from time to time.

Blowing the bellows to keep the fire glowing gave me heaps of pleasure. It depended on how much elbow grease I put into it to see the variation in the colours, and height of the flames. The pots were black due to the long use over the turf and wood fires. Coal coke was also used when it was available. The bastable and other cakes were delicious when freshly baked, especially when buttered with their own homemade country butter which was quite salty. Home-made blackberry jam was a delight and was easily made as there were plenty of berries available in the hedgerows around the farm. Crab apple jam was a great favourite.

Collecting eggs was one of my enjoyable pastimes. I went into the shed and cow stalls where the hens laid and I collected the warm eggs in my hands - about half dozen eggs at a time - and then carefully took them back to the kitchen. During one holiday I went to the cow stall, collected the eggs, and instead of walking carefully I ran straight out the door and into a heap of cow dung about fifteen inches high! I didn't see it on my way into the stall. The shock and shame of it! I was up to my knees in steaming hot, fresh, cow dung, which had been swept into a heap that morning. I smelled to high heaven, but didn't let go of the eggs. I knew I would never live it down, and all in the house kept reminding me to watch out for dung heaps in teases and taunts that went on for the duration of the holidays.

My brother Anthony and I had our first and last milking chore on the farm. When the cows were ready to be milked I took the end cow, whilst he took the second cow down from mine. We drew up the three legged stools, and sat down and then squirted a little milk to moisten the teat, as we had seen our uncle and aunts doing. They used the buckets to collect the warm milk and there was lovely foam at the top of the bucket as the milk flowed quickly. We decided to squirt milk at each other, under the cows' bellies. We both started to tug on the cows' teats and milk started to flow, usually into our faces, from both directions. We found this hilarious until suddenly, my cow got fed up because I must have tugged too hard on her, and she gave a big moo and kicked out. I got a fair belt from her and it stopped my messing straight away.

In those early days, my grandfather used to go into town at weekends for his pints. Normally, he went alone on the horse and cart. There was no car or other transport on the farm. He used to spend hours in there chatting to the locals and enjoying himself. When my brother Anthony won the all Ireland boxing title in 1940, at four stone four ounces, aged just ten years old, my grandfather took him to the pub and boasted about his boxing champion grandson. Of course granddad got plenty of free pints and boy did he milk it! We used to wait up for him to come home. He always did but was usually asleep on the cart after his fill of pints and Dolly, the mare, knew her way home. She used to safely deliver him after each trip, none the worse for his visit to the pub.

During those summer holidays in Kanturk I always looked forward with excitement to going to Mass every Sunday in the pony and trap. The trap was cleaned for the occasion and we all piled into it and listened to the clip-clop of Dolly's hooves as she trotted along on the country road in the beaming sunshine. Sometimes, she lifted her tail and left off wind, which made us duck down laughing in case anything else was discharged. We met lots of farmers at Mass, and grandfather bragged about his grandchildren.

One of the most important days in the summer was when the hay was being cut and saved. It was a most enjoyable period for all the children who watched all the activities under continuous sunshine. We enjoyed watching the threshing machine going full belt when the corn was ready for saving. In the early days the binding was done by hand, until machinery took over. As I watched the tractors going all out, and steam engines chugging away, I felt exhilarated. It was a treat when I saw my grandfather cutting the hay, using Jacko the stallion, and Dolly the mare, to pull the mower. My aunts came out to the fields where the action was taking place and brought with them strong tea, homemade brown bread, bastable cake, and some bottles of Guinness for the men. My cousins, brother and sister, and I got homemade lemon juice, made from fresh lemons. I thought how lucky we all were to be in this environment and how most of our school friends would never have the opportunity to savour all of this.

During the drawing in of the hay I went out with everybody and helped. They gave me a two-prong fork and I gathered loose hay and built it into a mini haycock. I was also allowed to stand on the float and spread the hay as it was thrown up onto it. The float is a flat cart without sides. Gradually, the hay built up and as it rose I got higher and higher on it. Usually, I had a two-prong fork to distribute the hay evenly across the float. Once, when I was just about to finish levelling the hay on the float, Dolly, who was hitched to the front, moved and caused me to tumble off the top of the haystack. As I fell to the ground, I saw the twin spikes of the fork come towards my side and for a split second it scared the life out of me. Fortunately, it went under my arm and slipped sideways away from me. There was general pandemonium and when there was hay on the float I was never allowed up on it again.

Usually, during the saving of the hay, all the farmers helped each other in rotation. Machinery was shared and manual labour freely exchanged. When the men returned to the house for dinner it was something to behold. A big, scrubbed, wooden table was piled high with lovely floury potatoes, home-made country butter, salted pork, home-made black and white pudding, sausages, Guinness, tea and more.

The threshing was a great day and the chugging of the threshing machine was music to the ears. The farmers forked the sheaves into the machine and the grains were separated from the sheaves. Chaff flew all over the place, particularly on a windy day, and it covered hair, clothes and equipment. The farmers used 'sugans' around the legs of their pants to stop the rats from seeking refuge where they should not. The 'sugan' is a rope made from straw and we all tried to learn the art of making this rope.

One day, my grandfather and my uncle Paddy, said they were going up to another part of the farm and asked me if I wanted to come along. I jumped at the chance. They hitched Jacko up and I was put sitting in the middle of the open cart, directly behind the horse, whilst grandfather was on one side, with uncle Paddy on the other. They kept smiling and talking as we went along the road. Next minute, I saw Jacko's tail go up and a loud explosion of stinking gas hit me in the face. Before I could do anything there followed a heap of hot manure, which landed in lumps on my lap. Too late I found out I had been set up. They laughed all that day, and the next. All the family did likewise as they were regaled with the story after I arrived back covered in horse manure. This was now my second baptism of dung and manure.

Usually, when I rode Jacko to the haggard he just walked quietly. But, one day, as I was on my way there something startled the horse and he bolted. I only had a muzzle rope on him, so this did not give great control. As he galloped, I got frightened and actually threw myself off his back into the ditch. Fortunately, I was unhurt, but I can still see Jacko's back hooves fly past my head. I mounted him again after that and all was well.

The thing I remember most vividly during my holidays in Kanturk was the slaughter of pigs on the farm. The cart was brought into the yard, near the back door of the house. The knives were sharpened, and water was boiled by the gallon. Broken glass was collected and a big bowl was put near the cart. Ropes were on hand, and everybody wore hessian sackcloth as aprons. The pig had been fattened, and corralled, and was brought snorting, squealing, and struggling to the cart. My grandfather and uncle Paddy then got the pig up on to the cart, turned him on his back, and tied one hind leg to the back corner of the cart. The front two legs were tied to the front corners of the cart. The pig was now secured, with his head forward off the cart, which was tipping towards the ground. The sharpened

BEYOND THE SEA

1936; Grandad John Regan

1936; Paal, Kanturk. Uncle Paddy and Dolly the mare, Auntie Nono, Nora,Eileen, Sheila,Mammy, Mary Anthony,and Auntie Kate.

St. Colman's Altar Servers' Society, Cobh.
— 16TH. OCTOBER, 1937 —

Back Row—J. Lynch (H. Row), J. MacCarthy, M. Smith, D. O'Halloran, S. Higgins, T. Higgins, T. Lynch, M. MacCarthy, L. Allen, F. Keating, T. Cahill, D. O'Kane, R. Hamilton, J. Halley.

Middle Row—M. Halligan, H. Roche, C. Ellis, G. O'Donovan, M. O'Sullivan, P. Coakley, M. O'Driscoll, D. MacCarthy, E. Stafford, J. Murphy, J. Keating, C. James, J. J. O'Sullivan, T. O'Brien, E. Cullinane, W. Murphy.

Front Row—J. Lynch (E. Hill), J. Chandler, P. Neville, M. English, M. Ronan, P. O'Mahony (Prefect), J. Walsh, S. O'Mahony (Senior Altar Server) T. Walsh (Prefect), M. O'Donovan (Prefect), L. Smith, J. O'Donovan, B. Neville, R. Kelly, G. Smith.

1937; Cobh Cathedral Altar boys

knife was drawn down the pig's throat towards the mouth. This initial cut was only to cut the skin and mark where the full incision would go. The squeals were unbelievable and the poor pig was terrified. Next, the knife was thrust deep into the pig's throat and bought towards the mouth. As the jugular vein was severed, blood gushed out and into a bowl, brought to collect the blood, which would later be used to make black pudding. When the pig finally ceased to struggle, and his free leg stopped kicking, the blood stopped flowing. The pig was now ready to be shaved. My aunts got the broken glass pieces and with boiling water shaved all the bristles off the carcass. The entrails were removed and all edible parts saved. The pig was now cut into various pieces, cleaned, and salted in barrels. All utensils, and the cart, were then cleaned and normality returned. There were some great days, and some sad days, but on the whole I would not change anything. Those holidays were the best I can remember.

Chapter five

Once every five years or so, the tarring of the roads took place. First of all tar barrels were placed at allocated spots on the road. A horse drawn tarring machine then came along the road and with the aid of a block and tackle the tar barrel was hoisted over a horizontal cylindrical type oven. When the temperature melted the tar sufficiently, it was manually sprayed over the road with a hose and spray unit. The men who used the equipment were heavily protected and covered in clothes and hessian bags were wrapped around their boots and up to their knees. The smell of hot tar was almost like a drug. I followed the tarring all the way up the road. Whilst the tar was still molten another crew came behind with gravel which they spread across the molten-tarred road, using hand shovels. It took days for the road to settle and I carried tar on my shoes all over the house. My hands usually got a fair share of tar as well. The way I got rid of the tar was to rub butter on it and eventually this got rid of the marks. If petrol or paraffin was available I used this. The tar melted every hot summer's day and I left the imprints of my shoes on it. More butter! It was easier to get than petrol.

Rubbish was collected in galvanized bins which had lids. The dustmen,-as we called the guys who collected the bins, used to hoist the bins on their shoulders and tip them into an open lorry. The bins made an awful racket as they were thrown back onto the kerb.

Lorries were a relatively new introduction then. Barry's Lorries delivered Stout to Mackey's and Jarmie's pubs on Harbour Row. Barry's were located at East Beach. The first lorries had solid rubber tyres. I used to stay and watch the wooden barrels being rolled into the pubs. When I sneaked a look into Mackey's pub, next door to us, there was sawdust on the floor and spittoons scattered for men to spit into. This was a common, horrible, unhygienic practice which occurred in all bars. There was very little light and the smell of sweat and stout was overpowering. Added to this was cigarette and pipe smoke, and this disgusted me at the time.

Women frequented the snug, which was a special little segregated, totally-enclosed, part of the pub. The women sneaked in for their half pint of Guinness or a tot of sherry or whiskey. Most, if not all, of the pubs were like this. One bright spot was that Mr. Mackey loved canaries, and had three cages with these lovely birds singing in the pub. Mr. Mackey usually asked me to collect grounsel for his canaries. I had no trouble finding this little weed locally and apparently the birds loved it.

Like all 'street kids' we had our own headquarters, or meeting place. It was Kidney's strand, better known to us as 'Port Sugar.' We named the strand 'Port Sugar,' because from a broken sewer pipe close by raw sewage was discharged and used to float on the water, where we swam. The term 'Sugar' was a local colloquialism for effluent. The gate to the strand is opposite the Preaching House steps. I use the term 'street kids' loosely, as we were really only friends who stuck together and played on the streets All the children, boys and girls, would swim together when the tide came in under the Baths' Bridge, into the strand. Suddenly, excrement appeared from the strand and from the adjacent broken sewage pipe. We thought nothing of pushing the brown lumps to one side and I still wonder how we never swallowed any of these foreign bodies.

Sewage was not the only hazard to be found here. Drowned cats and dogs and rats were frequently seen on the strand. Fido, a Jack Russell terrier and great ratter, once chased a rat that went into a hole in the wall on the strand. The rat's tail was just visible and Fido was going mad trying to grab it. Eventually, he got the rat and killed it, while we watched unafraid. The remains were left on the strand to rot. Nothing fazed any of us. We merrily carried on enjoying ourselves. I reckon all that exposure to dangerous and unhygienic materials actually helped to create good immune systems in us children; we never seemed to get anything worse than a cold.

Playing down at the strand was a favourite pastime. We used to race 'yachts' against each other and could spend hours whiling away the time down in this haven. To make 'yachts' we got bottle corks, a match-stick for a mast, a piece of paper as a sail, and a piece of slate, or glass as a keel. This was a great source of fun watching the little 'yachts' sailing out under the bridge arch, into the Harbour, to go where the breeze took them.

We were pretty much left alone there as children and learned to fend for ourselves. Unlike today our parents didn't have the time to watch over us every minute and we did sometimes get into trouble but we learned to rescue ourselves. Once, aged ten, I was practising my swimming during full tide when I accidently went outside my depth, by about one foot. I was directly under the arch and suddenly I realised how far out I was and panicked. I tried to catch a shackle that was fixed to the sidewall of the arch, the shackle was about four inches above my arm's length, and as I stretched, I missed and immediately went under the water. Nobody apparently noticed. When I reached the bottom I had the sense to hold my breath and started to walk under the water. However, I soon realised I was going out, away from shore, instead of in towards the slipway. I turned and kicked upwards to the surface. My lungs were bursting, air was escaping in large bubbles, and I was feeling faint. I let out the remainder of air as I came up. Once I broke the surface I gasped for air, and shouted, which only made me take in more dirty salty water before going down again. I kept my composure and I did not panic this time but kicked upwards and towards the shore. When I resurfaced my nose was just about over the water, with my feet on stones and I struggled ashore. The episode frightened me, but did not deter me from swimming there again. Neither did it affect my desire to go to sea years later.

Daddy used to worry about our antics at times but often took us to Cork on the train, where he treated us to cream cakes and lemonade, called Tanora, in the Tivoli Restaurant, in Patrick's Street. Then he used to take us for Polar ice cream in the South Mall. We loved this day out. The ice cream was twice the normal size, but when I look back it was not creamy like ice-cream is today. It was just iced milk but at the time it was great value and we loved it.

One thing that frightened me when I first went on trains as a young lad was the noise as the train approached Fota station. Sections of humped backed iron bridge rose up and then down, up and down and the hollow sound of steel over the bridge made me shiver. I could not understand why this should be. I thought there was some kind of monster outside the train. I remember ducking, and hiding, and it took a few trips for me to get used to this odd happening.

As a father Daddy could be very strict. We all felt his hand, or the strap, at some time. Occasionally, he had spells of bad temper and he acted impulsively. Three instances come to mind. Once was when he cleared out the pantry to build a shower. I was the recipient of his impatience. He passed three heavy books to me and one slipped to the floor. He picked it up and threw it back to me and it hit me on the nose, drawing blood. He immediately repented and calmed down. He must have had something on his mind, because he did not always react like this.

Another instance of his temper was when I heard Anthony getting a beating. He was upstairs in the top bedroom and Daddy was laying into him. I ran up and was just in time to see daddy with his arm over his head, holding an electric flex cable, ready to bring it down on Anthony's back. I immediately grabbed the flex, from behind daddy's back, and stopped him hitting Anthony. I was expecting a rebuke, or punishment, but daddy just lowered his head and left the room. That flex was never again used in the house. I never did ask, or was told, why Anthony was being punished.

Eileen also felt the outburst of his temper. He once caught her mitching, or langing, from school and broke an umbrella across her back. Langing was a Cork colloquial expression for bunking off school. Sometimes we deserved to be chastised, but other times he got it wrong. Anyway, it did not do us any harm. Within our family, nobody held grudges.

Amongst the many great memories I have of my childhood are the times that the Circus came to town. Duffy's Circus was a regular visitor to our town. The big parade was led by a number of elephants and clowns, followed by horses and caged lions, tigers, and other jungle animals. Music blared, and it seemed all the children in town followed the circus. The tents were pitched up in O'Reilly's field, at 'Top O' the Hill.' They were packed out for all performances, and the clowns were exceptional. The smell of the animals, and their droppings, outside the big tent was really strong. The noise of screaming monkeys, and the roaring of lions and tigers was terrifying when we walked close to the cages. When we sauntered amongst these cages it was nearly as good as going into the circus. Sometimes, we paid to get in, and other times we tried to sneak in under the canvas.

Other interesting memories include watching O'Reilly's horse drawn, two-wheel breadbox which was a common sight around town, as fresh bread was delivered daily. This van was beautifully decorated, with the name 'O'Reilly's Bread' painted on the sides. The driver sat on an open seat, on the front of this box. The seat was high up and he was exposed to the elements at all times. Bread had to be delivered at all costs. Thompson's bakery also supplied bread. Wooden racks were used to deliver bread from the van to the shop. There was no protective covering on the bread so the wonderful smell of fresh bread wafted out and made my mouth water. I loved the 'heel,' which is the crusty end piece of the bread. The more burned and crusty it was, the better I liked it.

The most common form of transport for getting around the town then was a bicycle. Bicycles had carbide lamps, and there was a foul smell from them. As far as I can remember, a piece of carbide was immersed in water and an inflammable gas was created and lit. With the aid of a reflector, behind a glass window, it provided enough light for the cyclist to be seen. Jarvey cars and horse carts had these lanterns pre-electric power. The Jarvey cars and Side cars or Jaunting cars, were the horse drawn, two-wheeled vehicles used to transport people before taxis became available. In the film 'The Quiet Man,' John

Wayne and Maureen O'Hara sat in one with Barry Fitzgerald acting as their Chaperone.

Down in West Beach, near the White Star offices, the Jarveys lined up with the Jaunting cars. There were up to six at any one time and all the owners used to stand around chatting, smoking, and drumming up business; particularly when liners and trains arrived. Close by, was a public toilet for men and this smelled to high heaven.

Three of the Jarveys that I remember were owned by Michael Halligan, Joe Twomey, and Walter Barry. The Halligan and Twomey families also had an undertaking business and their horse drawn hearses were common sights around town. The black horses were beautifully groomed and had their hooves blackened for the occasion. Later, Barrys became undertakers too and the three families also ran taxi services.

Further along by the camber there was a steel horse trough and this had a small trough low down at its side for dogs. Often both types of animals drank water side by side to quench their thirst. Nearby, was the mobile coffee hut where people gathered on regatta days and on Sundays. It was beautifully painted and maintained and it was where people were able to buy non-alcoholic drinks, sweets, and biscuits. Apparently the reason that it was on wheels was to avoid paying taxes, as it was not classed as a fixed building. It was moved backwards and forwards about six inches each year to avoid paying these taxes. In the camber nearby there were many boats and one named "Kathleen" was owned by Ruby Robinson and he used to take supplies out to the lighthouse. Many locals got free trips from him.

Chapter six

Besides enjoying myself all the family still had to obey the laws of the Catholic Church. As I got older I began realise and understand how the Church ruled our lives. Church laws defined how and when sex should be carried out and we accepted that, to a certain degree, without question. The laws of the church in this country have, for hundreds of years, governed both the private and public lives of its people and the issue of sex was one of the most fiercely governed. According to the Catholic Church, then and even now, sex was not for pleasure, but for procreation. This meant natural birth control was the only accepted practice during the so called, 'safe period,' in a woman's cycle. According to the church, sex for pleasure was wrong and had to be avoided, under pain of mortal sin, unless it was carried out as church laws defined it.

    The churching of women who had given birth was given by the Church, to mothers who had recovered from childbirth. It was a form of purification and thanksgiving for the woman who had recovered from bearing a child. The woman had to be catholic and married first off. She had to assure the church that her baby was not baptised outside the Roman Catholic Church. It was not classed as a precept, but as a pious and praiseworthy custom. At the end of the ceremony the woman usually left a half-crown, (two shillings and sixpence, old money) on the altar rails, in thanksgiving. It was always accepted that that any service performed by the church should be paid for, even if it was not compulsory. The shame attached to not paying was not worth the cost to pride, even to poor people. The church preached for large families and then charged the women for hearing the message. For some reason, a woman back then did not attend her own baby's baptism and christening. I think this too was a church law. Baptism was mandatory in the church, to cleanse the soul of new infants. The sin that needed to be expunged was 'Original sin,' which had been committed by Adam and Eve, when Adam ate the forbidden fruit. If a person died without being baptised then, according to the Church, their souls would go to 'Limbo' and not directly to heaven. I believed all these things growing up, there was no question of not believing them, and the power of the Church was so total. Later in life concerns about what happened to the babies' souls that were not baptised troubled me greatly. .

The half-crown that was the usual charge for a baptism in those days was a lot of money, compared to the wages earned. My siblings and I were born in the late 1920's and through the 1930's during the world depression. Money was scarce. In old money, a half-crown was worth thirty pence. You could get into the cinema for four pence, get a pack of ten Players cigarettes for six pence, five woodbines for two pence, a bag of ten sweets cost a half penny, and a pint of warm Guinness cost six pence. So the cost of purifying a woman's soul was five packs of Player's cigarettes or five pints of Guinness

In my time babies were baptised at the back of the church, not the front. This was also where the bodies of the faithful were left to repose in a mortuary after their death; they remained at the back of the church, and were not given the dignity to be allowed up to the altar. Mass and services for the dead were performed in the Mortuary. Both of my parents suffered this indignity. To the best of my recollection the corpses of the clergy were the only ones who were allowed up to the altar where religious services were performed for them. The body of the church is where the people practised their faith. I suppose, it had to do with the original design of the Church. Back in the old days, the Church was untouchable, and its word was law. There are catholic cemeteries, still in use today, where unbaptised babies were buried in un-consecrated ground. I don't know if these graves have now been consecrated. These differences, between those given the honour and dignity of consecration and respect and those not, were common throughout the system. As a family, when we went to mass we were separated across the aisle. Men occupied the left aisles and women occupied the right aisles. Thank God this nonsense no longer continues and today we live more like Christ showed us. In those days women had to have their heads covered when entering the church and men had bare heads.

There were a lot of holy pictures around homes in those days, as there were statues of the Blessed Virgin, Infant of Prague, Sacred Heart and St. Patrick. At the front door there was a little Holy Water holder and we had to dip our finger in it and make the sign of the cross as we entered, and left, the house. At Easter, we got the specially blessed Holy Water from the church. Sprigs of palm were in vases during Easter too. Most houses had firm religious beliefs and convictions. Children grew up with a sense of security and Christian love. Little oil lamps with crosses constantly burned in most houses in honour of the Sacred Heart. Daddy used to splash holy water over us, for any reason, as there was plenty of it available and it was free of charge.

The Eucharistic Procession was held in Cobh every year in June. This was the big outdoor religious ceremony of the year. We prepared long and hard for it and we collected all kinds of bunting, banners, statues, holy pictures, flowers, jam jars, plus any coloured paper we could lay our hands on to decorate the house for the occasion. The walls of the road were whitewashed in honour of the day and all kinds of greenery were tied to lampposts and railings to beautify the streets, especially on the route of the procession. Street windows were turned into grottos and altars and banners were strung across the roads. Our banner read 'Jesus, be not to me a judge, but a Saviour.' We filled jam jars with wild flowers and green weeds and these were placed every ten feet or so along the footpath. All in all, the whole presentation was sublime. Our street was one of the main routes for the procession and we were always proud of its appearance.

Usually, Benediction took place at the Bench; which was the large junction between four streets and overlooked the Harbour. During the procession, all the congregation split up into their own groups; Confraternity, Legion of Mary, Pioneer Total Abstinence, School classes, First Communion, Confirmation and general public. The Bishop, Archdeacons, Canons and

Priests came from the whole Diocese of Cloyne and the altar boys and Nuns were there in full regalia. Amongst these groups were various brass and reed bands and bagpipes. It was difficult at times to follow which hymn to sing, as each band had its own programme and they did not always combine for the same hymn and if they did they were out of time with each other. Amongst the Pioneer Total Abstinence group it was not unusual to see a few well-known characters who knew most of the local pubs, inside and out. We laughed a lot when we saw them but they ignored the banter. Usually, the day was sunny and everybody was uplifted at the end of it all. On the few occasions, due to weather conditions when it was necessary to cancel the outdoor procession, it took place around the cathedral grounds.

    I was an altar boy in St. Colman's cathedral for a number of years in the late 1930's. One evening during October devotions, amidst the waft of incense and the sound of the organ playing, I was situated at the side alter, where hundreds of candles were alight in candelabras. These were spread on the altar and on the floor. Suddenly, I felt heat and found that the beautiful lace on my Alb (Surplice) sleeve was on fire. I quenched it quickly but there was a bit of confusion for a while in the congregation. It meant I had to have repairs to my Alb. The Alb or Surplice was white, with laced design on both sleeves, and was worn over the cassock. As far as I knew a lady named Mrs. Keating used to do the lacework for all the priests and altar boys. One of her sons was an altar boy with me and another was in my class at the National school, nicknamed 'The Nash.'

    One day, whilst playing in front of my house, I unexpectedly got a message from one of the lead altar boys that Fr. Tim Murphy, the priest in charge of the altar boys, wanted to see both me and Denis O'Kane. Denis and I were close friends, and neighbours. When we got to the sacristy the Priest took a bamboo cane from under his cassock and accused us of eating nuts in the church during evening devotion. He ordered us to take down our trousers to receive punishment for this alleged offence. I deny to this day that we were eating anything and I believe it was the Bishop's two spinster relatives who reported us. Fr. Tim Murphy whacked us anyway when we refused to take our trousers down and we ran out of the Sacristy with him chasing us. Daddy was fuming when I told him and he went up straightaway to the Priest and in no uncertain language told him what he thought of him, and also told him that I would not be returning to serve under him again. Neither Denis, nor I, returned to serve Mass again.

Outside of being an altar boy I got up for eight o'clock mass each weekday morning with the family and we washed in cold water in the backyard, winter and summer, to prepare for it. I ran up to mass, rushed home for breakfast, and then rushed back uphill to school for nine am. The school was about a half mile from our house and was at the 'Top O' the Hill.' Cobh is a town of many hills and 'Top O the Hill' is a well known local area. It was tough going and I could not dodge going to mass because daddy always attended, and he could see me in the church. Sometimes, for evening devotions, I went to the church and waited at the back until daddy arrived. I made sure he saw me and then I'd scarper out, to rush down to the cinema to see my favourite films. The cinema filled quickly so I had to be early, otherwise I'd miss the show.

I had to attend evening devotions on a regular basis. These included Sunday Compline, which is the last service of the day in the church. Men's Confraternity was on Tuesday. Daddy was Prefect and I was unilaterally promoted his second in command, without being asked and without opposition. We wore Green ribbons around our necks with a medal attached. The Prefect wore a red ribbon. I kept the attendance records and diligently marked them in a book each week. This book was returned to the sacristy after devotions. Each street had its own section, so each individual was familiar to me. They would be too embarrassed to miss attending without good reason as those who missed would inevitably be observed by the some eagle-eyed parishioners who noticed their absence. The women held their Confraternity on Mondays. The 'Holy Hour' was held every first Friday of each month in the church and at home. There were various feast days, novenas, and annual mission visits which had to be observed. Weekly confession was a necessary obligation. There were the 'Stations of the Cross' and we had to do these on a regular basis, particularly around October and November as far as I remember.

We also had special appeals for the Black Babies, as poor children in Africa were termed then. We adopted them by the hundreds, by paying into collections and savings. There were regular appeals by missions for funds. Children in school contributed their fair share. Things haven't changed in that respect and I recall my own children doing similar charity work in their school days.

When the Missions arrived in town stalls and tents were erected outside the church and business boomed. Holy pictures, scapulars, rosary beads, prayer books and statues were sold incessantly. It was also a week of frightening sermons about Hell, Damnation, Fire and Brimstone. We were petrified when the Redemptorists and the Jesuits, gave their sermons.

They bellowed, "Keep your hands outside the blankets and across your chest in bed, avoid temptation and bad thoughts," Talk about being brainwashed! At one of these ceremonies, when the priest was in full flow shouting, and thumping the pulpit, he stopped and pointed at the congregation, in my general direction and he bellowed; "I mean you." He continued, "Hell and damnation awaits you."

I'm sure we all thought the same thing - he's pointing at me. Can he read my thoughts? It sent shivers up my spine. The man was serious! I wonder what he shouted at the women. The Missioners used to go around town, during devotions, and round up any male or female, who was not at the retreat. One of the missioners apparently approached a well-known local Protestant, named Ruby Robinson, and asked him why he was not in the church and Ruby allegedly said, "Because I'm not a Roman Catholic, I'm a Black Protestant."

A lot of the emphasis in religion then was on sex, for all its perceived badness they talked about it a lot! Other subjects were preached too but most of the teachings were done by fear and the threat of damnation forever. For instance, if you missed

Mass on Sunday, or Holy day, you committed a 'MORTAL' sin, and if you died then, without making a good confession, your immortal soul would go forever into the fires of Hell. Bad thoughts too, were enough to condemn your soul down to 'Old Nick.' As children there was no question of us ever doubting this possible outcome, hell was a very real threat and this belief still nags me to this day.

Recently, a priest preached that this method of teaching was wrong, and this comes as a relief to me, but I still have nagging doubts. For instance, I will not knowingly or willingly miss mass on Sundays, or Holy days, because I feel it is a Mortal sin. I will not go to communion, if I have eaten food less than one hour before receiving the Host. I think these are stupid man made laws, but it has been drilled into me and it's hard to go against teachings that are so ingrained in you from an early age.

In those days most people suffered from coughs and colds and it was during church services that this was most obvious. One cough during mass or during a sermon started an uncontrollable crescendo of coughing throughout the congregation. Sometimes it went on for quite a while and it was impossible to concentrate on the sermons.

After years of these lectures, when I began to develop a natural curiosity about the opposite sex, I began to worry about impurity. What about a kiss with a girl? Would that be enough to damn me forever, if it made me excited and I took pleasure out of it? What full-blooded young person can control these emotions? My first kiss was with a Cobh girl named Eileen, who lived up near the Preaching House. She was the first girl I dated and I was walking her home one night when I had my first kiss. We went up the Preaching House steps and stopped at the entrance to the gate. I knew I was going to try and kiss her and after a few moments of awkward silence I asked her, "Can I kiss you".

"No." She replied.

"OK" I said, and turned to go.

She quickly said, "All right, you can."

I kissed her awkwardly, once, and left. I didn't go out with her again, or with any Cobh girl. I fancied quite a number but was too shy to try. Most of the girls I went out with were from Cork. I enjoyed the kisses, and hugs, but being a 'good boy' I confessed any which gave me pleasure. Being normal, they all did, and I did not stop enjoying myself!

I still find it difficult to come to terms with some of the teachings and commandments of the Catholic Church, as they have changed so much over the years and so much of what I have learned as a boy has now changed since then. More things have become accepted as normal by the Church, i.e. relaxing of fasting and abstinence. Indulgences, as I knew them, are virtually gone by the board. It was forbidden to eat solid food for twelve hours before receiving communion when I was a child. Now we have Saturday evening Mass, priests and nuns in ordinary civilian clothes, lots of changes. As for indulgences, we were taught that you could get plenary or partial indulgences, if you carried out certain religious duties and prayed for the Pope's intentions. A plenary indulgence, or partial indulgence, was used for the souls of the departed, whereby they could be released from Purgatory when you prayed for them. On 'All Souls' day you gained a plenary indulgence, when you visited the church, nominated the soul, and prayed for the Pope's intentions. You also had to have made a good confession and communion. Nowadays, all this is changed and plenary indulgences are more difficult to get and I wonder what happens to the Poor Souls in Purgatory? I think, but am not sure due to all the changes, that some form of partial indulgence is still available. I don't hear of indulgences these days. Come to think of it; if I met all the requirements of the plenary indulgence for a particular soul surely I didn't need to do it again the following year for the same soul, but I continued to rescue the same soul every time I got a chance. The souls might not have been in Purgatory in the first place. Gone are the days when priests used to walk on church grounds reciting their 'Office.' in public.

One priest used to cycle around lonely roads and the 'Bush Field' where he hoped to catch courting couples, and on discovery he would beat them with sticks. The usual give away for the couple was their two bikes together, left somewhere near where they were courting, while they enjoyed themselves. Rumours abounded about how long it took the priest to chase the couples, after first discovering them. The bishop did not allow priests to have their own cars. All their movements were on foot, by bike, or by jaunting car. This was not great for anyone who needed spiritual help, or the Last Rites, and could sometimes end up waiting long periods of time for the local priest to arrive. The priests may have had some access to local taxis.

As well as church collections, the 'Stations' were another way of getting money in. This collection was made annually for the support of the clergy. A list of streets was read from the pulpit and a day was designated for householders to come to the Church and pay up. The priest waited at a table at the altar-rails. He had his book containing all the names in it and when the person from the house went to the altar-rails, in front of everybody, they handed over their offering, which was duly noted in the ledger. The following Sunday, the amounts donated were read from the pulpit. The person's name was read out, with the amount paid. The priest started with the largest amount paid per person, until he got down to poor people who could only donate pence. It degraded and embarrassed these parishioners. There were also Sunday collections at the door, usually under the watchful eye of the Parish Priest, so nobody could dodge contributing.

With all the money coming into the church from contributions, donations and indulgences the Catholic Church was one of the wealthiest institutions in the country. Outside of the church the rest of the country suffered greatly from lack of upkeep due to lack of funds. Footpaths which took us to the most beautifully kept churches were uneven and badly repaired, and roads were more like dirt tracks.

Chapter seven

It was approximately 1932 when I started school. My earliest memory of school is of meeting Sister Agnes and Sister Gabriel in the convent run by the Mercy Nuns. I don't remember whether or not I cried on my first day. Our class was called the 'Babies' Class. At that young age we were mainly taught music, singing, and colouring. We also had little toys to play with and the usual squabbles took place over 'ownership' of particular toys. For our music lesson Sister Agnes issued the class with sticks about six inches long. Attached to each end of these sticks were small bells which we shook furiously to create sound. There were some tambourines and drums also given to some children. We all wanted the tambourines or drums, but alas, I never got one to show off my musical talents! That time, whenever a child was deemed to deserve punishment Sister Agnes would give the child a six-inch ruler and tell him or her to administer a token smack to his or her own knuckles. It never hurt since even at that tender age we were not stupid enough to hit ourselves too hard. There was a large rocking horse in the classroom and rides were given to all the children except me. Sister Agnes and Sister Gabriel called me to one side early on and explained that my father had told them that I had a slight rupture (hernia), and should not be put on the horse. I was bitterly disappointed but accepted the decision. I used a truss for the hernia and it eventually cured itself without surgery. I remember I still had it years later, when I was boxing. However, there were plenty of other amusements for us children and we played happily with each other. There were the occasional rows, at which time the ruler would be offered to the offender, to self-administer the relevant punishment. They were extremely happy days for me and I was sad leaving that school.

   Sister Gabriel always smelled of carbolic soap and had a clean, scrubbed look. Even her glasses were spotless. Sister Agnes was older, smaller, and was not as impressive as Sister Gabriel. Sister Gabriel used to glide into the classroom. Both were kind, gentle ladies, and it's strange how I have these specific memories of them. Perhaps, I had a soft spot for Sister Gabriel. I went back to see them once, when I was working but I regret that I did not try harder to go and visit these lovely nuns before they died. The two nuns were always together. I hope someday that they will be together again and greet me at the 'pearly gates'.

The convent was a mixed school of boys and girls. We had the same play breaks, and when the boys rushed to the outside loo; which always had a horrible smell from it, some of the girls giggled and watched us, and the boys used to try and piddle on them. The girls thought this was very funny and laughed as they ran away.

1934 to 1940 were my informative school years. From the convent, daddy took me to St. Joseph's National School, which was a boys' school. It was christened 'The Nash' and was known locally by this term of endearment. It was close to the Convent, at 'Top O' the Hill.' The regular routine in St. Joseph's was that during breaks the boys were free to play in the school yard and when the whistle sounded we stopped playing and waited for the second blow of the whistle, at which time we ran and formed into lines for our own particular class. Our teacher waited for us. When a line was formed silence had to be maintained or else we got a wallop if caught talking. We were then led back into our class. Throughout the years, this discipline was maintained.

In those years the schoolyard was constructed from hardened clay. In wet weather there were many mud patches. There was no shelter from the rain and more often than not we missed our play time outside. Later, the yard was concreted and a shed was built at the one end.

Here I met my first male teacher, Mr. Casey. I started in lower infants under Mr. Casey's tuition. We used chalk to write on slates. It was in later classes that we had paper and ink. The ink was made from powder, mixed with water. The inkwells had to be replenished on a regular basis and Gerard Bransfield used to make the ink. We used various types of nibs and we all had our own favourite nib. I liked the slim type and hated the fat stubby type. Nibs had to be changed regularly, when they became corroded. I don't remember this practice ever stopping at the 'Nash.' Ceramic inkwells were used for the ink and most were chipped and had stains on them. Sometimes, boys spat into the inkwells and this became obvious when the pen was dipped and it came out with a sticky mess on the end.

Later, the slates were used for another purpose. The slate was fitted in a wooden frame and Mr. Casey removed one end of the frame so that the slate could be moved in and out of the frame; like opening a sash-window. He used this when we were preparing for our first confession. He sat down and we knelt in front of him. He moved the slate through the frame in front of him, as if we were in a confessional, and we faced him and started to tell our 'sins,' practising for our first holy confession. We had to make up sins then, innocent as we were, so that we could go through the pretence of making a confession.

In preparation for Communion we were taught Catechism. This was a religious lesson that was learned from a little blue book that cost a few pennies. Each child had to have a copy of it and had to learn the contents off by heart. A caning was administered if you incorrectly answered any question. The teachers wanted to show off their classes to the priest who came to visit. The priests would ask us questions from the book and they too were not very kind if we failed to answer the questions correctly. One Parish Priest used to take the class for catechism examination, and the first thing he did was to go to the cupboard for the cane. He used to hold the cane across the back of his neck, with both hands, bending it from time to time as he walked up, and down, between rows of desks, asking questions. If the wrong answer was given the cane would descend with venom across the boy's back and shoulders. The priest got his fix this way. He was known as Buck Jones. At that time chastisement by teachers and priests was common and we had no recourse but to accept this punishment. Naturally, none of us liked it, particularly as some punishments were over the top and undeserved. Even now I can still see in my mind's eye the cane descending. I'm sure it helped us to remember our lessons and spellings. I remember how we learned to spell catechism through a mnemonic. It was: Cathy Athy Told Eddie Connell How I Stole Matches.

When I made my first communion it was great fun. I went around my relatives and neighbours, telling them I had made my communion. The whole idea was to collect as much money as possible, and to get sweets and cakes. Communion was secondary. However, I felt like a big boy going up to communion each week in church after that. One thing still mystifies me, and that was how I committed so many sins, that I had to go to confession every week. I'm sure I committed more sins by making them up, and confessing these to the priest, than all the sins I actually committed. Most of my sins were, 'being bold at home,' 'telling lies,' and 'I can't remember any more.' That's another lie!

Once, when I was going home from school I noticed a pigeon's nest in the side of a building. The bird flew away when I approached with some friends. The nest was in full view and had five blue eggs in it. I threw a pebble and broke one of the eggs and immediately one of the other boys said, "Lynch, you committed a mortal sin." This terrified me and I was scared to tell the priest in confession. It took me a long time to get over this and I wondered how many other mortal sins I committed by not confessing it immediately at my next confession. I dared not think where I would go if I died at that time. Just imagine now how one boy can terrify another, by telling him that an act committed was a mortal sin. These teachings hold powerful grips on a young person's mind….

I remember that the poorer boys seemed to be treated more harshly than other middle class boys. This was a trend I noticed throughout most of the early years, up to sixth class. For instance, on a number of occasions a particular teacher of younger boys, took a boy and wrapped brown paper around his body and head, and he then tied a string around the paper and pretended to put a stamp on the paper, to post the boy to China. The boy was then pulled crying to the boiler room and left there for some time before being released. It used to terrify these little boys. This was only done to satisfy the teacher's lust for power, not to teach us anything. Physical punishment was administered just because boys misbehaved, did not learn their lessons, or simply that the teacher did not like them. The smallest infraction could lead to severe physical punishment. We were only small children. I also saw one teacher throw an old penny coin at boys to get their attention. We did not have the same fun in this school as we did at the convent when we were smaller. During later years, a number of selected boys had to take some of the 'slower' boys to one side and help them with their weaker subjects, whilst the teacher would carry on teaching the rest of the class. In some cases it was not easy, as some fellows did not appear to want to learn. During this period I missed out on the new lesson being taught whilst I was teaching the other boy and had to do my own catching up.

    One of the teachers I clearly remember from my younger days was Mr. Cooney. This man was a real gentleman and he taught me how to write longhand. I remember that the lines on the pages for writing were coloured red and blue. The red lines were to use for capital letters, while the blue lines were to help you write the lower case letters. I don't remember Mr. Cooney ever lifting a hand, or cane, to any boy. Later, he lodged with my sister Kathleen, and many years after that I was to see him again, though this time at his funeral in Dublin. He was a kind and lovely man and I have fond memories of him

    Cheap meals were given every day to children from poorer families. The cost was one penny. These meals were given out in a small hall, near the back of the Cathedral and they were known as the 'Penny Dinners' or 'St. Anthony's Bread'. A lot of the better off boys tried to sneak in for a meal. I never did because much as I wanted to do so, I felt too ashamed or afraid in case I got caught. It was a different matter in school when current buns were delivered to the school, to be handed out to the poorer children, for midday break. When these children got their share, there were usually some buns left over and the teacher would ask, "Who would like a bun?" Every single child's hand would shoot up. The teacher broke the buns into pieces and shared them out. Unfortunately, he did not do a 'Sermon on the Mount' miracle and only a few children got some so there were lots of disappointed faces. However, this was rectified over the week, as each boy got a piece of the bun at some stage.

We had great fun around the cathedral corner on the way to school when it was a windy day. The wind blew very strongly and we caught the tail ends of our coats, raised them above our heads, behind our backs, and they ballooned in the wind. We then ran against the wind. Sometimes we were blown back up against the rails, which were about eight feet above the road below. We could easily have been blown down onto the road. It was so exciting, and we often spent an hour or so playing there on the way home from school. On a fine day I used to see priests, dressed in their cassocks, walking up and down the side of the cathedral whilst they read their breviary. A number of times I sprained my ankle running down the steps from the cathedral to the road. This was because I tried to take two or three steps at a time. I often wonder now why I continued doing this as it was quite painful and sore, perhaps it was just the invincibility of youth.

At lunchtime, most of the boys ran home at twelve thirty for lunch and had to return again for one thirty. For some like me, it was all downhill on the way home but uphill after consuming a meal. It was no wonder I didn't have an ounce of fat on me. For mid-breaks I normally had bread and jam. Others had bread, butter and sugar sprinkled on top.

My favourite Brother teacher was Cormac. From what I remember he taught me through fourth and fifth class. He was a tough taskmaster and used to bind the end of his bamboo cane with waxed hemp. He administered a hefty caning when necessary and even though I got one or two whacks, I still respected him a lot as his punishments weren't unfairly meted out. He was the teacher I had when war broke out in September 1939. He used to leave the class to get regular news bulletins about the impending war. When the news broke about war being declared Brother Cormac came into the class looking very thoughtful, and said, "Boys, war has been declared by

Britain on Germany." We did very little in school that day as teachers were meeting, and discussing, the prospects for all involved. I, for one, did not take much notice of the conflict that was brewing. It was too far away. Everything here appeared normal. Later, I followed it very closely in the newspaper, The Daily Express, which gave maps with British and Nazi stickers pinpointing the war zones.

Another teacher in the 'Nash' was Spud Murphy, who was a small dumpy man with a nasty habit of blowing the contents of his nose directly on the floor, and then he got one of the boys to clean it up. One particular boy suffered this ignominy most of the time. Thank God I never had him. I don't think I would have put up with that kind of treatment. A number of mothers got physical with him over various years, because of his aggression and bad manners. He hid in the press whenever Mrs. Cashman came gunning for him. When he got married everybody was gob smacked. In fairness his wife could have done much, much better for herself. There were plenty of jokes and stories going around the town at the time.

The school had a lot of teachers that I either was taught by or knew of during my time there. Mr. Casey was my first teacher. He was OK but frightened some of 'slower' small boys. Mr. Cooney was a gentleman and it was he who taught me to read and write. Brother Anslym was my Irish teacher and he was very good. Brother Lyness had a reputation for brutality, thankfully he never taught me. Brother Ugenus was Head Brother at the school, he was a fair man and most people found him fine. Then there was Mr. Sullivan, who was nice enough but was a stickler for pronunciation of the English language. He never taught me. He did however remind me that the 'Statute of Kilkenny' was made in the year 1367, and I have never forgotten. This happened when he asked me for the registration number of daddy's Baby Ford car. I said, 'ZB 1367' and this is how I was reminded about the Statute being signed in Kilkenny, in AD 1367.

Anyway, back to school. Cruelty was evident in some cases, for instance some of the boys were treated terribly and for various minor infractions had serious physical punishment as a consequence. Boys were caught and dragged by the ears. Some were lifted by the sideburns. Once, a boy had a tuning fork lodged in his neck after the teacher threw it at him. Sometimes, blackboard dusters or coins were thrown at us to request attention. The dusters were pieces of wood with a felt pad attached and really hurt on impact.

We were threatened with being sent to Greenmount Industrial School regularly. This scared us as this particular institution was where unruly and naughty children were sent to be held until they were old enough to be released. It had a reputation of being a very cruel and uncompromising place and lately the news stories are full of the horrors that took place at those types of institutions.

Throughout our school days boys were punched and beaten with canes, with particular aim taken to ensure the cane caught the thumb or wrist. I occasionally suffered this experience, and it was painful. Back then, all subjects were taught through Irish. We learned Algebra, Geometry, Logarithms, Vectors and fractions even in the Irish language.

When boys had squabbles and differences in school, it usually ended in a challenge to a fight. In class a boy looked straight at his opponent and made indications what he was going to do to him, up in the quarry after school. A fist to one eye, indicated a punch to the eye, and then the other replied indicating he was going to blacken both eyes. This posturing continued and indicated the punishment about to be inflicted on his opponent. Each fellow's pals arranged the time, and agreed how long the fight would last. Normally, nothing happened except a challenge to strike the first blow. Each one declined to strike first and both saved face in more ways than one. Some good fights did take place however, but not necessarily in the quarry. These were usually spontaneous, and there was no time to think before somebody acted. I had a few spats in the schoolyard, and in one scrap I knocked two front teeth out of a fellow named Flynn. It was more by accident than design, but I got total respect after that, and was a school hero.

Other fights went on in the schoolyard during breaks. I remember one that was taking place when the Head Brother, Ugenus, arrived. He stopped the fight and then told all the boys to form a circle around the two pugilists. Next, he said he would referee the fight, which was to be fought under the Marquis of Queensbury rules. And so it did, for six one-minute rounds. We all enjoyed it, and so did the Brother. Fortunately, only occasional blows were landed, and neither boy was hurt. The fight was declared a draw.

During my school years we got up to many capers and pranks. Many boys played truant, and there was a policeman, Guard Cahill, known locally as 'Skatum', who kept an eye out for such carryings on, and checked the attendance roll every week. We treated him with great respect and fear, even though I never remember him being a rough character.

On occasion, after a particularly bad fight a boy might have to visit the doctor. Dr. Ledlie was a female doctor who acted as medical consultant for the boxing club. The boys felt embarrassed when she asked to check if they had hernias. Fortunately for me I never had reason to go to her. I did over a period of time attend other doctors. Dr. Hegerty once examined me before I applied for the Metropolitan Police and made the comment;

"Boy, you have the heart of a horse and legs like tree trunks."

Nurse Collins was the district nurse. She brought most of us into the world. Sister Finbarr of the local Bons Secours Nuns also did lots of health checks amongst the local people and was very well liked, and respected.

One day at home a frightening thing happened, and it was lucky daddy was in the house at the time. I happened to go out in the backyard and there were two small legs sticking up and kicking in a full barrel of water. This barrel was used to collect rainwater from the roof, via a broken gutter through a down pipe. My brother Anthony had climbed up, to look in the barrel and had fallen headfirst into it. I shouted for help and tried to pull him out of the barrel but couldn't manage it until luckily daddy arrived and lifted him out. Anthony was lucky that I happened to go out in the back at the time, otherwise he would certainly have drowned, and none of us knew how long he had been submersed in the water. Thank God, he came around amid much spluttering. Daddy got an axe and immediately broke the barrel into smithereens.

During these years and when we were quite young, after school we used to visit wakes in any part of town that was holding one. It was mainly curiosity which drove us. We used to say a few prayers as we looked at the brown habits worn by dead men, light blue habits for women, and white for babies and young children. Sometimes, a prayer book was placed under the chin of the deceased to keep the mouth shut and occasionally a penny covered an open eye. Other times a ribbon was tied around the head and chin, to keep the mouth closed. We did not seem to mind and after the first wake we got used to it and never had much fear of the deceased.

One particular wake sticks in my mind in the 1930's. It was the wake of the late Bishop of Cloyne, Bishop Robert Browne. We were fascinated; because it was not always we got the chance to see a Bishop laid out in full regalia. We were told by our teachers to go and pray for the Bishop and we did not need a second command. The Bishop was laid out in the Bishop's Palace and we all stayed for quite a while, but I think we did more chatting, and looking around us, than praying. The Bishop had pink slippers on his feet and these were the focal point for a number of boys, who kept touching them. One young boy, however, could not contain himself, and removed one slipper and promptly ran away with it, leaving the poor Bishop minus one slipper. Some of the bigger boys caught up with him and replaced the slipper on the Bishop's foot. We all then made a hasty retreat lest we get in trouble.

Later, I attended the funeral of the bishop, which took place in St. Colman's cemetery. I believe that he is the only Bishop of Cloyne who was not buried within the church grounds. I did not attend the funeral, or the wake, of the late Bishop Roche, RIP. This particular Bishop used to strut up and down the streets, in all his finery, and held his hand in front of him so that all the faithful could kiss his ring. We ran towards him, to kiss the ring, and the adults did likewise. It was expected that we perform this ritual and the Bishop loved it. I don't remember seeing any signs of acknowledgement or humility from the man. When I think of this practice of brainwashing it makes me realise how conditioned we were. After devotions one winter's evening, it was quite dark, and daddy stopped outside the church but he was still within the grounds of the church. He took out his cigarettes and then he lit up one, while talking to a friend.

A voice behind him sternly said, "Put that cigarette out."

Daddy turned around and started to say, "Mind your Bloody…" when he saw it was Bishop Roche.

"Sorry, your Lordship, My apologies your Lordship," said daddy, whilst kissing the ring.

Oh! The shame of being caught smoking inside church grounds. Daddy was a member of the third Order of St. Francis, Total Abstinence Pioneers, and prefect of the Confraternity. He carried out collections at the front of the church, and now this! He got over it.

One day, as I walked up Harbour Hill, I saw Jack Doyle, heavyweight boxer and singer, known as the Gorgeous Gael. He was young, impressive looking, and very well dressed. A large number of local children and teenagers followed him around and they all looked very happy. I saw him again later, when he was married to Movita, the film actress. He was quite a sight to see and very famous in our area, and in the whole country because he had boxed for the British Heavyweight championship.

While at school, I continued to collect bicycle parts and I added any unused surplus parts to the heap of scrap iron, copper, and lead I had made to make a few bob for the boys. Since there were a number of us involved, we shared any money we made with our joint efforts. Once a week a 'Scrap man' came down from Cork, in a battered old lorry, and collected scrap. We patiently waited for him, and when he passed up Harbour Row, on his way to Kidney's pub, we stopped him and offered up our scrap, which he weighed by lifting it and using the weight of it to estimate its value. To get a few more ounces of weight we used to fill the gas lead pipe scrap with sand and we then hammered the two ends together to hold the sand in. Maybe he was wise to this and said nothing as he was also paid by the weight of it all when he returned to the yard in Cork. We got our few pence and then we followed the lorry to the Bench. We waited, outside Kidney's pub, and when we felt that the driver was settled, we sneaked up and took some scrap from the lorry. This we stored until the following week. We got away with this little scheme for some time before he caught us. He still took our scrap every week, but he kept an eye on his lorry.

My father was always very good at mending and fixing things and with a rake of young children running around all the time breaking things his skills often came in handy. Once, a group of boys were playing outside our house and Seamus O'Kane was sitting on the windowsill. He was three years older than me. Daddy started messing around with us and poor Seamus crashed through one pane of window glass. The window was made up of three panes, and was now short a pane. Fortunately, Seamus did not get injured. Daddy had to put up a temporary sheet of plywood until he had the glass replaced. Trust daddy to do it on the cheap again. This time he got a large mirror from some shop. The dimension lengthways was about one foot short but he had a plan for this. The width was too wide but this could be cut to size. Not only did the silver mirror coating on the back of the glass have to be removed but there was also a hole in the centre of the mirror, which had to be filled in. The hole was about one and half inches in diameter. This was no problem for daddy and he glued a piece of picture glass to cover the hole. The main work again fell on me, and I had to scrape away the silver coating of the mirror, and I can tell you it was not easy, especially as the scraper was small and blunt. Having measured the window properly, daddy fitted a piece of wood as a frame at the top of the mirror pane. Next, he fitted a new piece of glass to fill the gap. Here we were now with two full panes of glass, and one pane with a hole in the centre, and a dividing frame about a foot from the top.

Precautions should now be taken to avert such a catastrophe happening again. Again, another of his brilliant ideas came to him. The stair rods were substantial, and round, so he removed them and went to a Blacksmith, named Mr Allen. This man made a guardrail for the front window and used the stair rods as vertical rails. Daddy took great pride installing this brainwave of his.

A great place for passing the time was at Poole's billiard and snooker room on East Beach and this was a smoke filled place where lads went on the mitch from school, and spent many evenings playing both games. This shop later became a Fish and Chip shop.

There were two hardware stores that I can remember from the early years. Madigan's had a very large store in the square but it burned down. Cronin's Barrett's had a store in the Street known as 'the Beach', and sold bicycles. The owner was Dickie Bah.

It was in the Bath's Quay, previously known as the Atlantic Quay that I saw daddy's first car, and I sat on his lap as he practised driving, in and out between the benches which were spread around the quay. The quay was a large recreation area, like a park, and was covered in gravel. It must have been about 1934 or thereabouts as I was very young. I can still visualise a Morris Cowley open saloon car, grey in colour, with no hood, and it had a circular glass steam pressure valve with a needle, indicating pressure, on the front of the bonnet. There were lots of little levers and knobs on the wooden dashboard and on the large steering wheel. The bonnet was in two halves, and opened from the sides, where it was secured with leather straps. The wheels were thin and narrow. I loved that old car but I don't remember going anywhere in it, except down in the Bath's Quay.

His next car was a Baby Austin. I think the windows were made of mica or similar material. From what I can remember, it was a flexible scaly type of yellowish material, which could be peeled off in layers. We sat inside and did not get wet because it had a hood. It was painted a dark colour, which may have been brown. I remember I used to pick off pieces of this material when nobody was looking.

Daddy then bought a Baby Ford, and like all Fords it was black. It had red leather upholstery inside. Soon after he bought this car daddy gave a lift to a farmer, who accidentally cut the back seat, and it now bore the sign of a shoemaker's stitch. She was a lovely little machine, which took daddy and Anthony to Dublin and Clonmel for the Boxing tournaments. This was the car carrying the number plate ZB 1367. I loved going on trips in it and felt like it was my car. This car was shining black due to my elbow grease and daddy insisted I was responsible for maintaining it in that condition. He gave me a tin of wax polish, which had the name Simoniz on the tin, and which was very difficult to work with, but under constant supervision and hard graft, I produced a shiny mirror type finish. Daddy loved to brag about reaching sixty mph on the Curragh road when he went to the Boxing championships in Dublin.

Daddy used to wear a half plate of dentures but he never felt comfortable wearing them, though he would always smile to show them off when he was in public. It was a common routine for him to wear them as he drove in town, waving to people, but the moment he reached the outskirts, out came the dentures to be dumped into his coat pocket. They were retrieved and replaced in his mouth when he hit town again. He very rarely wore them in the house.

He also wore glasses for reading. Whenever he got these glasses changed for himself he passed the old ones on to mammy as she needed reading glasses too, but never got tested or ever owned suitable glasses. Poor mammy put up with it and lied to say the glasses were fine, but I could see by the way she kept trying to adjust them that the glasses were not in any way suitable.

The car was garaged in Lynch's Quay, in a double garage, adjoining the Social Welfare centre. I used to go and get it started for daddy whenever he went to use it. I loved doing this but on reflection I had to turn the engine by hand, so maybe daddy was not being kind but just being clever by getting me to start it as it was a difficult job. Later, as I got older, I changed wheels, tyres, and found small faults, and rectified them.

This was the car I remember that we went out in most times. My friend Frank Harris used to wait and see if any of my sisters dropped out, so that he could get a trip in it. He loved the car and later in life told me how much he envied us. We were often dropped to school in the car and felt very important. Unfortunately, World War II, our Emergency, happened and the car was laid up in Paddy Barry's farm in Ballywilliam, in Cobh, for the duration of hostilities. After the Emergency daddy could not get the car to start and sold it cheaply to Jack English, a garage owner, in the Bath's Quay. On reflection, I remember daddy sat in the car and tried to start it, but I cannot remember if he charged the battery, the spark plugs, or even put petrol in it at the time.

Even later, he got a Ford five-CWT van. This jalopy never seemed to have any go in it, and struggled up the smallest incline. It was great at going downhill though! I think daddy's intention was to use it to move stuff and make a few shillings. It could hardly move itself, and never did make a penny for him.

Outside Jack English's repair garage, down in the Bath's Quay, close to Kidney's Strand wall, there was an old hand-operated petrol pump. Customers had to pump the petrol themselves by using a hand lever. When the lever was pushed backwards and forwards, a one-gallon glass jar at the top of the pump filled with petrol. When full, and when the lever was released, the petrol flowed into the car tank. This was repeated for each gallon of petrol needed.

Adjoining English's garage there were about twelve single attached garages, built for private car owners who rented them. I don't know who rented them or who the owner of the garages was. It may have been Bath's Company.

I watched as a car was ready to be serviced in the English's garage one day. It was up on the 'H' shaped hydraulic lift. There appeared to be nobody around and I got too inquisitive. I saw a lever and curiosity got the better of me so I pulled it. Disaster! First of all, I had not noticed a bicycle parked close to the lifting ramp and secondly, the car started to descend on the lift. I saw that the wheel of the bike was going to be damaged, but I was glued to the spot, consumed with panic and I forgot to reset the lever. Crunch! The front wheel of the bike was slowly mangled between concrete and metal. Ouch! A startled mechanic appeared, as if by magic and ran to restart the machine.

"What's your name? What's your father's name?" the mechanic spat out at me. I told him and he made a note of it. He continued, "Where do you live?" Again, I told him. He advised me in no uncertain language where to go, and that my father would take care of me.

My intention was to get home before him and to be first to tell daddy the news, and hope he would be lenient and understanding. Shamefacedly, I went home and admitted what I had done and daddy was quite fair. He bought a new wheel, with the proviso that I had to repay the money when I started work. I don't remember if he ever deducted it from the money I handed up when I did start doing part-time work.

I was fortunate that Alfie Packham, who worked as a shoemaker at Lena May Connors, used to ride a lovely pony into town from Ballymore. Poor Alfie had a disability and needed the pony. I got the job of riding the pony from Harbour Row, up Harbour Hill, Thomas Street, and into Dan O'Connell's forge in the Quarry. Dan took care of the pony until evening, when the pony was taken back to Alfie.

The forge was close by the school and we often went there to watch the blacksmith working. He was well protected from the sparks and fire. His clothing was heavy and he wore a leather apron. It was fairly dark inside the forge and the clinkers from the fire were crunchy underfoot. There was an acrid smell in the air. This was due to the hot metal horseshoe being fitted to the horse's hoof. On contact there was a sizzling sound and smoke rose as the shoe was fitted. Before the fitting, the blacksmith removed any old shoes from the animal, and pared the hoof, to ensure that there was a good fit.

The fire was on a raised open hearth and was fed often with coal. This was raised to white heat with the bellows. The tools used were an anvil, hammer, knife, file, and tongs to handle the hot metal. A large bowl of water was kept for cooling the hot metal. The metal was heated until it was white hot, and then it was beaten into shape on the anvil. This beating, and shaping, created lots of hot metal sparks which flew around. There was also a lovely musical rhythm as the hammer hit the metal. Extreme caution was needed at this time. The shoe or any other metal object that was cooled in the water would release sizzles of steam. I liked to watch as cart wheels were repaired, or new metal wheel-bands were made and fitted.

Occasionally, I went to the local slaughterhouse after school. I saw all types of farm animals being slaughtered in those early years. Bulls were hauled in with a rope tied to the rings in their noses. The rope was fed through another ring in the concrete floor of the slaughterhouse and on to a hand controlled winch machine, which pulled the bull in until he came down on his front knees, with his head at floor level. The fatal sledgehammer blow was delivered between his eyes, and then a sharp knife was stuck into his throat until the jugular vein was severed, and his lifeblood drained from him. As far as I know no humane stun-guns were available in those days and it was all very brutal.

Sheep had two front legs, and one back leg, tied together and there were up to half a dozen sheep at a time tied like this awaiting slaughter. One by one, they were pulled in and the butcher put the sheep's head between his legs and stuck a pointed sharp knife into the neck of the sheep until it came out the other side. He then twisted the knife until the jugular vein was severed and the blood flowed freely from the wound. All this time other sheep watched this ongoing butchery. When pigs were slaughtered, it was similar to the operation in my grandfather's farm in Kanturk, except there was no cart used.

I took the pig's bladder and used it for a football. Sometimes, I'd let the bladder dry out, before I blew and tied it, but other times I blew them up straight from the slaughtered animal, blood and all. I don't know why, but this butchery did not affect me then as much as it did in later lifetime.

It must have been around 1938 that I started to watch Cove Ramblers playing soccer, up in McCarthy's field (now the GAA pitch at Carrignafoy.). Some of the players I remember from the first games I watched are; John Coveney, Denis Harrington, John Cotter, David Broderick, Mick Doherty, Baller Wall, P. Burke, Moss Cummins, Fanty Keating, Jimmy O Mahoney, L. Pinkney, Willie O' Keeffe and Jack Barry.

The Wanderers also had a team managed by Pake Hurley and there was great rivalry between both these teams. When Wanderer's went bankrupt most of these played for Ramblers. I followed Ramblers to Cork and had great days at the Mardyke Flower Lodge and Turner's Cross.

It was in the late thirties when my father experimented with headphones, and a 'cat's whisker,' as the crystal radio was then known. The 'cat's whisker' was only a piece of wire and this delicately touched a spot on a silicon crystal, to give a directional flow to the signal. He was thrilled when he picked up Radio Eireann for a while. Apparently he heard the Radio Eireann introduction music of 'Proudly the note on the trumpet is sounding...'

He passed the headphones to me to listen, but no matter how hard I tried, I could not hear a thing. I couldn't understand what all the fuss was about. Later, I understood the complexities of trying to get the 'cat's whisker' aligned with the one spot on the crystal, which gave the signal flow. This was to be the forerunner of the crystal diode, transistor, and the microchip. It was sometime later before I realised how involved I was to get in the use, manufacture, and testing of these components. However, this was a long time off.

In the meantime, lots of families listened to the battery operated Wireless. The main programmes listened to were; Question Time, along with GAA matches. Radio Eireann did not broadcast a full day's programme. We particularly liked the advertisements, especially those for Imco dry cleaning, and Urney chocolate.

One of the big events on radio was when the draw took place for 'The Irish Sweepstake,' the forerunner to what we now know as the Lotto. Everybody hoped they would be amongst the lucky ones. Tickets were sold everywhere and were even smuggled into the USA where they were illegal. Daddy bought a share at every 'Grand National' race. These could be bought as a full ticket for one pound, or a half, quarter, or one eighth shares.

To operate the Wireless to hear the programmes it was necessary to have good supplies of electricity. This was supplied by a wet cell battery, which consisted of lead plates immersed in sulphuric acid, and these cells were enclosed in a glass jar. A dry nine-volt grid bias battery and a dry 120 volt high-tension battery were also needed. This meant that all these sets were operated by direct current. The arrival of electricity on stream changed all that. All that was now necessary was to connect to the mains electricity.

Radio Luxembourg was a favourite for music programmes then. BBC and Radio Athlone were our principle stations. Radio stations of most countries were clearly marked on the dial. After Radio Eireann got the name, our station broadcasted from Athlone. It originally broadcasted from the GPO in Dublin. Now we had to twirl the various knobs to tune in a station. I did pick up German broadcasts during the war, with Lord Haw-Haw giving his spoofs. "Germany calling, Germany calling, this is station Bremen on the Reich…"

To receive good reception daddy and I fitted a long aerial out in the back. Daddy used a weight at the end of the copper wire aerial, and swung the copper wire and insulator up vertically to Harbour View. Two ceramic insulators were used, one at each end. A down lead copper wire was connected to the radio. Wireless was the name given to it in those days, but the term Radio was used later. The reception was usually very noisy, and crackly, due to local electrical interference and thunder, and lightening. Sometimes, signals faded and later stations began interfering with each other, due to bandwidths overlapping.

All in all these were very happy days and I learned a lot about life and its good and bad sides. School was never a chore and I vowed that the cruelty dished out by some teachers would not affect my admiration for the rest of the teachers. My home life and the townspeople in Cobh were a positive and great influence on me, and the pace of life was brilliant. People nowadays work nonstop and have not the time to fully enjoy life. There was still a lot to learn and cope with and I looked forward to meeting these challenges.

Chapter eight

Electricity seemed to be in our house since I was very young as I can remember some rooms that had brass switches and there was a candelabra in the sitting room. It's possible that some electricity may have been supplied by the old Power House down in the adjacent Baths Quay prior to the arrival of power from Ardnacrusha.

Growing up I worked in McIllwraith's electrical shop in East Beach on a casual basis for no pay, in order to learn about electrical wiring, battery charging, and other useful things. Anthony also worked there, much more often than I did. Mr McIllwraith rewired a lot of the houses in Cobh and used to service, and sell, wireless sets. The Fitzgerald's, too, had a shop on East Beach and they were also involved in this type of business.

East Beach, The Beach and West Beach are streets in Cobh. From the junction of Harbour Row and Harbour Hill the street which joins this junction is called East Beach and this joins The Beach where it links to West Beach. The whole series of buildings which occupy this street were originally built above the High Water mark of the adjacent beach before the British built walls, quays, slipways, and piers and reclaimed part of the beach for road-building, and to assist their military operations . The street is the main shopping centre in Cobh along with Harbour Row. Mr. McIllwraith rewired our own house for electricity soon after it was available in Cobh from the main National source. All the wiring was taken up on the surface of the walls, and sometimes the wires were stapled to the walls. Other times, wooden channels were used to encase the wires. However, at the beginning, only the rooms were wired for lighting and there were no lights wired for the hall and landings. The wiring for the landings and hall was carried out in the same way later on. Until then, we had to carry on going up the dark stairs at night, with lighted candles. A number of nasty things happened amid the joy of lights on command. The switches had brass covers, which gave a shock to unlucky ones from time to time, due to dampness and condensation. Likewise, the kitchen walls and water pipes were alive due to bad earthing and dampness. We got many shocks before the full dangers of electricity became known, and long before safeguards were implemented.

Light bulbs did not last too long. On regular occasions there would be a sudden flash and the bulb was kaput. When a fuse blew it was the usual practice to repair it, with silver paper from a cigarette packet, if the correct wire gauge was not available. At first we had no idea of what a dangerous practice this was, but we soon learned.

Up to this point we had only the solid fuel stove for our cooking and heating requirements. This meant lighting a fire and cleaning the ashes every day in order to keep us all warm and fed - a tiresome and messy chore. Daddy also had to go to the flue on the outside wall and light papers to make sure there was enough of a draught for the fire. It was a game of chance. Now, with the arrival of miraculous electricity Daddy went and bought an electric cooker and dismantled the old stove. He cemented and bricked up part of the open space and built a smaller fireplace instead. It was at ground level and was no more that about one foot wide and a foot deep. It was a miserable little fire but conserved fuel, which was the objective, now that we had the new cooker. The new cooker though, for all its modernity couldn't make toast and trying to toast a round of bread on the now tiny fire was hard work. A slice of bread was attached to a fork and held near the fire and we had to wait for the miserable little flames to do their work. Often I gave up.

When flashlights later became available with the increased popularity of batteries we had great fun on 'All Souls Night,' more commonly known as Halloween night. From a dark corner, behind bushes, or from behind a tree, we popped out in front of people, clutching a flashlight under our chins whilst pulling a gruesome face. The results were spectacularly hideous! Before the spooked people knew what had happened we would have disappeared, leaving them screaming with fear. We also got large turnips and hollowed them, carving faces out of them, and would place a lighted candle inside – slightly less sophisticated pumpkin. These even frightened us! In those days people genuinely believed in ghosts and spirits so our pranks were all the more satisfying to us!

Down the Beach, in the centre of the road at the Camber, there was a big electric light stanchion. This was a very high metal lamppost, with the light at the top. This was originally erected when electricity first lit up the town. There were others placed in the middle of junctions of the Bench, at the top of King's Street, and on Harbour Hill. There were others placed on the footpath at Harbour Row.

The street lighting lit up from the main source of power now meant that we could wander around town at night and see the different views of shops lit up with electric lights. Some of the displays in the windows looked different than they looked in daytime. One such shop was the local fish shop down town, in the Beach, where Molly Dorgan sold fish. Mackerel, cod, herrings and plaice all reflected the light. As the fish were all freshly caught that day the sight was wonderful. We used to sing a song about her and it went like this:

'Molly Dorgan sells fish,
Three ha'pence a dish,
If you don't like Molly Dorgan,
You won't like her fish.'

"They're in, they're breaking." The shouts were heard throughout the town. The summer months brought the mackerel into the harbour. The seagulls and gannets were the first indicators that we were about to enter a period of constant fishing, a period of productivity, plenty and enjoyment. These birds squabbled and fought as they dived into the shoals of fish throughout the harbour. Everybody; men, women, boys and girls, scurried down to the various slipways and quays, to cast their lines out and to get a good seating or standing location. It was frightening at times to see very young children as they sat on the quays with their feet dangling over the edge. Crowds converged on the town from all over the county, and further afield. All kinds of boats moved up and down the harbour as they followed the mackerel breaks. The sight of mackerel breaking, and the activity of birds and people on a sunny day, once seen, was never forgotten. From the deep-water quay, along the beach to Lynch's quay, across the Baths, and down to the Camber in the Holy Ground, there were hundreds of lines in the water. In the slipway at Lynch's Quay the water was thick with sprats, and the mackerel came in almost to the edge of the tidemark. The sun sparkled on them, and the speed at which the mackerel moved was unbelievable. One minute a shoal of mackerel, their green backs to the surface, went by, cruising, and swimming in formation. I remember this sight as a thrilling and beautiful memory. Suddenly, they came upon the sprats and they broke out of the water and travelled at very fast speeds, with silver streaks glistening. Injured sprats flapped around on the surface and were gobbled up by the frenzied mackerel.

Boys and girls waited on the slipway, knee deep in sprats, and when the mackerel followed the sprats, the kids all kicked together and dozens of mackerel and thousands of sprats jumped and flailed on the slipway. It was a feast for all but sad to say lots of the fish were left to die and were thrown back into the sea. Normally the conger eels and crabs feasted on these carcases.

With normal fishing, everybody pulled in the fish as fast as they put their lines into the water. We always seemed to catch lots and never had an empty line. Even silver paper was used as bait and worked. Sometimes, overexcitement caused amateurs to pull too quickly when they got a bite and the mackerel ended up overhead, dangling on the telegraph wires. Some unfortunate children and a few adults too, had to go to the doctor or chemist to remove hooks from different parts of their bodies. Kids went from house to house giving fish away because they had too many – it was a plentiful time - cats and dogs had the time of their lives. There was not a hungry animal in town during these days of silver enjoyment. Occasionally, some of the lucky fellows caught a Johnny Dory, which seemed to follow sprats. My clothes used to be regularly soaking wet and covered with fish scales but I never noticed and would wear the same clothes the following day. It was a common sight to see boys, girls and adults walking around with dried silvery fish scales on their shoes.

When I wanted a change I progressed to conger eel fishing. There were always eels where there were mackerel. Their targets were the mackerel. There were some special places where I could get good big congers. Usually, there was a hole or hiding place underwater where they congregated. I used to throw out mackerel heads and this brought the congers nearby very quickly. I fished for them in the Camber inlet located down the Holy Ground, in Lynch's Quay, and off the deep-water quay. Sometimes, I tried the Bath's quay. I caught them in all areas. Some were up to seven feet long and had vicious mouths of teeth, which I avoided at all costs. They could inflict a nasty wound, or could even bite the hand off you. When I got them to the edge of the quay, I jammed the gaff into their mouths to keep them shut. This a long stick with a metal hook for landing large fish. Then I avoided the thrashing body and removed the fishing hook. Next, I threw the conger over my shoulder with the gaff still attached, winding him as he landed on the quayside. I used to then club him to stun him and finally, with a knife, I severed the spinal cord. This was necessary, as I could not take any chances with this fish. I saw one, its head almost severed, grab a piece of three-inch wood and sink its teeth into it. It did not let go until it died.

Daddy used his initiative to create unique fishing bait. When he saw any horses left by farmers outside Mackey's' pub, he would go up to the horses and cut off some of the horse's tail hairs to make bait. He tied the hair to hooks, with beads which he pinched from my sister's necklaces. He had some great success with this and caught dogfish, pollock and mackerel at Lynch's Quay. Mackerel were suckers for any bait and virtually gave themselves up. Sometimes, he did not arrive home until after one a.m. and mammy often asked me to go and check that he had not fallen into the sea. When I would arrive down in the middle of the night to find him still fishing he would welcome me casually and tell me to sit down for a while. He would take a packet of Player's cigarettes from his pocket and light up. He didn't smoke much, and certainly was not addicted. When he finished the fag, he gathered his fishing rod and any fish and we'd head off home. It was usually about two a.m. by the time we returned.

Our next-door neighbours at No. 34 were the Barry family. There were five siblings living there; Charlie, Frank, and their sisters; Winnie, Chrissie, and a third whose name escapes me. Two of the sisters lived in America. One was secretary to Henry Ford and the other worked with very famous American actresses. Nobody in the family of five was married and all were heavy smokers. I heard them coughing and spluttering all the time. They had brown nicotine marks on their fingers and upper lips. Frank was the worst addict and when he coughed it was frightening.

He kept greyhounds in a shed at the back of his yard. This shed overlooked the roof of our shed. These animals created all kinds of noise and on one occasion they got out onto our shed. Since our shed roof was sloping corrugated iron they found it hard to hold their footing and kept howling, and barking until they were eventually shepherded back by Frank. He used to rub foul smelling stuff on the dogs to prevent, or cure, distemper. I hated that smell when they passed me. He used to boil sheep's heads for them too. The combination of smells from there was awful.

Frank was a Petty Officer in the Coastal Patrol during the war and was a survivor of the ill-fated disaster of the 'Irish Popular' in Cork Harbour, when a number of people were drowned. Charlie served tea, sandwiches, and other items on the tenders that which serviced the liners. My sister Sheila did cleaning for him on these trips but he never gave her one-penny for her efforts. She did not mind much though as she got trips to the liners which she loved. Apparently, Charlie sobbed and cried openly when Mammy died. He was very fond of her and brought her soup from time to time. Chrissie Barry also did Bed and Breakfast for people going to the USA. When they had an overflow they sometimes asked mammy to put up some of the emigrants with a bed and they would continue to supply the breakfast. In our house they only got a large bowl and a jug of cold water for their washing needs. Charlie opened a small library. The family also had a Shell petrol pump outside the house.

Anthony told me he was down in the Bath's Quay one day, when he looked back up towards our house. There was a young woman standing at the upstairs window upstairs in Barrys' house and she was completely nude and kept waving at him. She seemed to enjoy teasing him. Ask him how he felt - he certainly did not move from this vantage spot too quickly!

One day my sister Nora got a call from Miss Barry to say her sister had died. She asked Nora to assist her as friends were calling for the wake. Nora got into her best Sunday finery to help and greet the friends. When ready, she called next door to Miss Barry, where she was greeted with a welcome and told how good she was for offering to help.

"What would like me to do? How can I help?" Nora asked

Miss Barry replied, "Nora, would you polish the brass work and scrub the front door step for me?"

Nora was stunned and went back home, muttering under her breath, and changed back into her work clothes. It was far from prayers she was muttering under her breath! When she returned and completed the job Miss Barry searched around and produced a bottle of Gin and said to Nora; "Thanks, Nora, you've been very good. Now we'll have a little drink."

Nora told her that she did not drink spirits and Miss Barry squinted at the bottle and exclaimed; "Where did that come from?"

She then produced a bottle of sherry and they both had a drink. Nora felt enough was enough and said she was going home. Winnie asked her to come back later, to wake her up if she fell asleep as she had to meet somebody three hours later. Nora duly called in to the house and there was Winnie fast asleep with the Gin bottle lying nearby. Nora tried for a half hour to wake her up; pushing, shoving and shouting… but all to no avail. Eventually, Nora gave up, said "feck her" and returned home. When Nora told me this story I laughed and jokingly told Nora that she was too soft. However, The Barry family were good neighbours and often did favours for my family.

Often, we Lynches would have baths in Barrys as we did not have the luxury of a bath in our own house. One day daddy arranged for us all to go in next door for our baths. Seniority was invoked unfortunately for us little ones! First in was daddy, to lounge and soak in the clean soapy water, followed by mammy, into the same water. Eileen and I were next to share the by now slightly murky bath water. We were young then but I can still remember the water was tepid, almost cold, when we got into it. Four people bathed for the price of one - economy!

Growing up in Cobh one of the most important events of the year were the pre war Regattas held on the 15th August. They were never to be forgotten events. Returning immigrants on holiday and people from all over the county came to enjoy the festivities and fireworks. There was nothing to match that day throughout the year.

In the daytime it cost six pence to enter the Promenade in order to see and hear the bands playing in the lovely bandstand, and to watch various events, including the greasy pole. This was a long pole extended over the water from the adjacent pier. The pole was covered in grease, all the way from top to bottom. The challenge was for fellows to get on the pole at the pier and try to wriggle out as far as possible without falling into the sea. It was such a sight! Sometimes pillow fights took place too at this event and these were hilarious. It was great fun and every year the same fellows tried again, and again, with no success. Some fellows nearly lost their manhood as they gamely tried to get to the end of the pole. Girls also had a go with the same inevitable results – ending up in the cold sea, soaking and laughing as the rest of the town watched from the quayside.

From the vantage of the Promenade, we had a first rate view of the fireworks which were set off at ten p.m. on the Quay. It was also a great vantage point from which to watch boats and yachts racing. The four man plus coxswain crews in gigs from Whitepoint and Rushbrooke went head to head for honours.

Crews from the Bench and Carrigaloe also participated. Others who took part later were Crosshaven and St. Finbarrs. The races started at the Hulk, near Cuskinny, and ended up at the Promenade. We all had our own favourites and cheered them home. Whitepoint was my team. The Geary brothers of Carrigaloe were also very good too though. Other oared boats which raced were; yawls, punts, and paired whaleboats. Lots of locals took part in all types of races. The sea and every imaginable activity that could take part on it were such a large part of life in Cobh.

    During the regatta British Royal Navy warships and other vessels were festooned with flags and lights. In the evening searchlights blazed into the night sky from ships and Forts. Liners or other ships in port did their share of celebrating by decorating and blowing their hooters. These scenes were also repeated on New Year's Day. The British navy certainly knew how to set off a fireworks display. It was breathtaking. Royal Navy destroyers from England shipped all the fireworks in for the occasion. As part of the celebrations the regatta even had an annual Beauty Queen crowning. The first that I remember was Helen Halligan. She was the sister of Michael, who was in my class in school. They came from a family of undertakers. For the celebrations the town was decked out with bunting, flags and other decorations. Shops, stalls, and other vendors did a roaring trade then. There were bands and singing, as well as talent competitions all day long. The regatta brought all kinds of new people to Cobh. Sieks with their turban covered heads and leather suitcases used to come and sell their silks and tapestries during regatta time. We were quite frightened of them at the beginning, mainly because of their dark skin and beards but ultimately found them to be quiet people who just went about their business. 'Shawlees', Cork City women who normally traded on stalls in the Coal Quay in the City and who wore their black shawls over their heads, arrived in force in their charabancs. Soon they were in full bartering and bantering mood. Many of these older women dyed their hair black by using black ink and when it rained their faces were a mess of black streaks. They were good fun though.

The Tinkers fascinated us when they came around, mending pots, pans, and any utensils made of tin. They would do the work on the footpath and were excellent tradesmen. No such thing as begging with these lovely people. They worked hard for a living.

Jewish men also visited the town and daddy did business with one of them once during the insurance agents strike. When the strike happened and daddy couldn't work there were no wages coming into the house and though Daddy did not want to go on strike at the same time he did not want to be known as a 'blackleg.' A 'blackleg' was a union person who broke ranks and continued to work whilst his friends stayed out on strike. Anyway, he had to sell the chandelier, dining room suites, piano, briefcase, brass weighing scales, and various other items to the Jewish man to keep his family fed. This was around the depression of the thirties and a very hard time for everyone.

When we were very young we watched the regatta from the bedroom window but later, when we were old enough we enjoyed being part of it all down on the quay with the happy throngs of people. When we were older still and in our teenage years we particularly enjoyed the regatta for all the Cork girls who came to Cobh each year for the festival!

At the 1935 Regatta four hundred balloons were released into the sky. Each balloon was sold in advance and the buyer could attach their name and address to the balloon. Daddy bought some balloons for all of us and we excitedly released them. There was a prize for the one which was picked up furthest from Cobh. If memory serves me correctly, one got as far as Germany and another reached France! We didn't have any luck though.

Daddy's clever initiative often came to the fore around this time when he would brainstorm new ideas to profit from the festivities of the regatta. He owned a beautiful one metre long model yacht, which had Bermuda rig and one year he painted white with 'Valspar' paint. He now got the bright idea to get the company, 'Valspar', to sponsor the yacht in a race to be held in Cork Harbour on Regatta day. He even named the yacht 'Valspar'. Unfortunately his idea didn't float and he didn't get the sponsorship, and I'm not sure if he even got a reply. He was a cute and crafty man who always said; 'Nothing ventured, nothing gained'. As far as the race went I can only remember daddy tying the tiller, a precursor to the more modern autopilot, and the yacht going off in the opposite direction, with us frantically chasing it in a punt.

The arrival of the amusements at regatta time was one of the biggest joys for children, and indeed some adults. I loved the Bumpers and the Shooting Gallery' Sometimes, I went on the Chair O' Planes and had lots of fun catching the chair in front of me and kicking it forward and up. This was often done to me and it was thrilling. There were also the Swing Boats, which were great. I remember, one time, wearing out the seat out of my trousers from going on all the rides and having the embarrassing time of trying to get home unobserved, with a large hole in my trousers and no underwear.

During the daytime people enjoyed lovely recitals on the Cathedral Carillon, by Dr. Staff Gebruers, Snr. The Cork Butter Exchange Band was one of a number of bands which played regularly in the bandstand, in the Promenade, during the Regattas.

On some Regatta days a cycle race was held and this started at the Post Office on West Beach and finished back there after the cyclists had raced out the Lower Road and back via the High Road. It was very exciting as it was open to all to take part in. One winner I remember was Mr Dudley Balburnie - Bal as I knew him - who later guided me to my future career as a Marine Radio Officer.

Another gruelling cycle race also took place around this time. This race was an endurance test and the object was to find the cyclist who could climb the Barrack Hill on a bike in the fastest time. From a push start, at Frennets' cinema, the cyclist had to cycle to the top of this steep hill. It was breathtaking to watch one fellow after another as they tried to negotiate this enormous task. Some took it straight; others zigzagged to lessen the steepness of the hill. Few achieved it. One who did was Ronnie Twomey, a member of the Cove Boxing Club.

One very sad personal memory for my family is always tied up in those of Regatta day. It was the day before Regatta day that my baby brother, Jeremiah, died of cardiac failure, 14th August 1939. He was only fifteen months old. I can remember to this day how I found out;

Daddy's crying awakened me. I heard him call mammy and say; "Mammy, God has taken him from us." I got up and went down to the bedroom and there was this lovely little boy, only fifteen months old, at peace. We all cried together. We had a cocker spaniel dog, named Lucy at the time, and she would not come out from under little Jerry's bed until after the body was taken and buried. While Jerry was ill with the bronchitis that led to his death Lucy had stayed in his room with him round the clock. Later, the white coffin was taken to the new grave at Carrignafoy, where he was laid to rest. Mammy, daddy, Patty, Mary and now Jerry's niece baby Nora have since joined him. It is only about two hundred yards from the grave to the house where daddy and mammy first settled in when they got married, and where I was born.

In the late thirties the Cove Boxing Club was first opened down in the Beach, at the back of The Rainbow Shop adjoining Wilson's chemist shop. The Rainbow Shop window front area was curved and led into what was really only a long corridor, leading into a room at the end which became the Boxing Club. Sweets were sold in the shop/corridor and there was no way of closing off the shop area when the front door was opened. There was never a reported theft of anything from the shop. The founders of the Boxing Club were My father, Jack Lynch, who was timekeeper and secretary; Harry McCrossen, who was the trainer there; Eddie Nolan, who took on the role of treasurer and referee; Bertie Maguire, who was our local Garda Superintendent and Mr. Kavanagh who worked in Wilson's local chemist shop.

I, my brother and many of our friends were members of the club. I must confess that I never had a love of boxing and did not apply myself to it like others did. All I wanted to do was to get out of the ring as soon as the bout finished, win or lose.

I only did it because I was expected to do so by daddy. My brother Anthony and Chris Walsh were known as the 'Cove midgets' and boxed exhibition bouts in the City Hall in Cork, and other places, at four stone four ounces in weight. They were excellent and drew applause from everybody.

Anthony was the first Cobh boxer to win an all Ireland boxing title. In 1940 he won the title at four stone four ounces, when he was just ten years old. He also won many County and Munster titles, at various weights. Some years later he was runner up in the seven stone, seven ounces weight category for the Irish title. For being the best losing finalist of the championship he was awarded 'The Lombard Cup'. Our local paper, 'The paper,' said 'he was robbed' and we all agreed. When he was in the Royal Navy he won more boxing titles and continued to pursue his passion for the sport. After the untimely death of the trainer Harry McCrossen, Garda Eddie Smith came and trained the lads. He was ex-lightweight champion of Ireland. At a tournament held in the 'Coliseum' (Young Men's Society Hall) Garda Smith gave exhibition bouts with various members of the Cobh club.

When the club was relocated in the old Preaching House Guard Carberry tried to set up a Brass and Reed Band. I was one of a few boys to join. He taught me music terms like quavers and semi quavers but that's as far as I went with it - because in my whole time there I never got to blow an instrument or beat a drum because there were no musical instruments available. Suffice to say the venture soon ended.

Ronnie Twomey, who was a boxer in the club and he worked on improvements needed in the club. The committee decided that it was necessary to start a fund raising effort. One thing that was needed was a constant flow of cash to supplement the Flag Day, concerts, boxing tournaments and voluntary contributions. Card games such as Whist and '45' drives were top of the agenda but tables and chairs were needed for this and so first money raised had to be targeted for this objective. With a bit of luck and hard graft eventually enough was gathered to buy the tables and the bench-seats were made up from discarded floor joists. Elbow grease and dedication shone through as nails were removed from the joists. The wood was planed smooth, cut, and shaped as required. Ronnie did most of this work and soon the club had a fine layout for these card drives, which provided urgent funds. There were also many good supporters in Cobh and Cork City who gave silver cups, medals, and shields as prizes and trophies for our competitions. It was a community effort that got the club off the ground and functioning well.

There was a variety of characters that were part of the club then. The club had all types as members and each unique person brought something to the table, even if it was only just a laugh or a joke. Once, at a whist drive, Guard Cahill was sitting at the card table for the games to commence. He loved his card games and chatting up the ladies, particularly if they had sweets. This particular time, this lady got fed up with him as he never handed his sweets around so before she had come on this night she had doctored her own confectionary. She had got a bar of Fry's cream chocolate, split it in half, and replaced the sweet cream filling with white soft soap. At the whist drive she gave him a piece of the doctored chocolate which he duly gobbled up. As he started to chew all hell broke loose. The table of cards was overturned as he leapt up and down shouting and trying to spit the soapy chocolate out of his mouth. There was general hilarity and Guard Cahill certainly learned his lesson!

My love of boats was also a general love of transport of all kinds. Whether it was a train, car, side-car or one of the hundreds of types of boats in Cobh, to me if it moved it was fascinating. My friends and I used to stand on one of the bridges straddling the railway lines and wait for the train to pass underneath. It came along with smoke billowing from its funnel and made a hissing sound that only the old steam trains could produce. As it passed, the carriages swayed and the clicking noise of the wheels hitting the rails was quite audible. We coughed, spluttered, and heaved from the smoke but for some reason we did it over, and over again.

Outside the station, close to the Stationmaster's house on the Lower Road, there was a water tank that was used to replenish the boilers of the trains. Trains stopped here regularly and a canvas pipe was lowered to the train and the boilers were filled for the next journey back to Cork. Further out at Whitepoint there was a large table to turn the train and this fascinated us. We watched as the train was uncoupled at the station and then came out backwards from Cobh station where it stopped on the turn-table. Next, it was manhandled and turned before it returned to the carriages at the station. It then hooked up with the carriages and headed back for Cork.

Still with transport, and out on the water, the Morsecock was a two-funnelled boat that took people on excursions around the Harbour and normally anchored up off the Deepwater Quay. From time to time she acted as a tugboat to tow vessels. She was involved in s.s Celtic salvage operation. As well as this activity there was a constant coming and going of British destroyers and battleships, which were moored in the harbour. They could be seen launching their boats to let sailors get shore leave. The Spike Island launches, used for taking British troops backwards and forwards between Cobh, Haulbowline and Spike Island, were constantly moving across the harbour. These were also used for towing targets outside the harbour, where artillery guns practiced firing.

I watched Irish emigrants going to England on the Innisfallen and Kenmare. The pre-war Innisfallen was a well-known passenger ferry-boat that was sunk by the Germans. Her successor, also named Innisfallen, used to blow her horn as she came around the Spit lighthouse, homeward bound, and this was the signal for everybody to come out and wave, whether or not they had friends aboard. Two Captains of this ship were from Cobh; Captain Horne and Captain Hamilton. People looked out their windows and on the quayside waved handkerchiefs. Tears of sadness, outward bound, and joy when homeward bound were common.

At the Deepwater Quay, people cried and sang, "Come back to Erin" every time the Tender left with emigrants for the liners which were heading for America. Sometimes the singing was supported by a band, including the local Sandy Marshall's Fife and Drum band. I was too young to understand the loneliness they felt and I laughed at them with the rest of the lads. Usually the town was crowded with visitors for the liners. Later, I was to feel some of their loneliness when I left the town for England for the first time on the same Innisfallen that I'd waved to previously. Still later, when I served on the deep sea tug, m.v Turmoil I felt the same sadness when I saw Molly, baby Ann, my sister Sheila and my parents waving to me as I passed down river to leave the harbour on a salvage job.

When the coronation of King Edward VIII took place in 1938, a huge replica of a Crown all lit up was placed on top of the Spike Island. It was magnificent. Destroyers in the harbour flew all kinds of flags and searchlights lit up the sky. The ships sirens hooted and blared, as crowds cheered downtown. Little did they know then how long it would be before he abdicated.

Just before the war Laurel and Hardy came to visit Ireland in a very special trip and I followed them from the United States Hotel, now the Commodore, to the quay, where they embarked on the tender for the Liner. We were used to seeing these comedians in the cinema but to actually see them in the flesh was great. We cheered them and some tried to shake hands with them but not many succeeded. My luck was in as I actually touched Laurel's coat before I was ushered away. I stayed until they vanished into their cabins and were no longer visible. The Church carillon chimed out the famous signature music associated with them and they were overwhelmed that this was done for them.

Growing up then we got our education where we could and as for learning about sex that was knowledge that was even harder to come by. I remember foraging about in our larder once when I discovered two very attractive hard-cover books. One was an Encyclopaedia of World war one containing photos and a story of soldiers' exploits which intrigued me. The other one though was even more intriguing! This book was a mixed selection of stories, some historical, some detective stories but best of all, a romance! It was this story which gave me my first piece of sex education. The story was called 'Messalina, the Illustrious Harlot' and I read it over, and over again. One sentence remains etched in my memory forever, it went like this; 'In the morning, she pushed the naked body, still clinging to her, away.' Even now I can visualise myself, sitting at the window reading, and rereading, this story, and absorbing this new experience. Now I knew what a harlot was. I smile now when I think of our total ignorance of sexual matters and reflect on our idea of what we thought was a 'loose woman.' Believe it or not as kids we thought that it referred to a woman who might have had a strange way of walking until the adults taught us the new more shocking meaning.

 Apart from that shocking discovery life was relatively innocent. As I grew up daddy used to take myself and my siblings around with him in the car, whenever he was collecting insurance. Before he had a car, and during the war, he cycled around the districts on his bike. I remember feeling sorry for him when I saw him come home soaked to the skin, frozen with the cold. His area covered Cobh and the surrounding places in the Great Island, and into Carrigtwohill, Midleton and Watergrasshill. He collected in Spike Island and Haulbowline too, where British service men, and women, resided and held insurance policies. I got to know many of these people while I accompanied daddy to work.

   He used to take me around Midleton, East Ferry, and Cloyne; places that were his stomping grounds when he had been growing up. When I got older, and had my own transport, I frequented these places too. Often, he went down to the water's edge at Walterstown, across the river from East Ferry. He used to sit there on the wall as he reminisced and would point across the water and say; "Up there, on that hill, and over there by the beautiful East Ferry church, I used to walk when I was young". Straight in front, at East Ferry, was Murphy's Pub and this was another focal point for him. It was where he used to have his pint when he was young and single. He also said that he used to sit on the wall at Aghada to watch the Cobh regatta fireworks and occasionally he came across on the Ferry, which ran from East Ferry to Walterstown. He then walked into Cobh and back again later at night. Apparently, a whistle was enough then to attract the ferryman, who rowed across to collect his passengers. He charged two pence per passenger and six pence for a bike.

    Daddy made good use of his contacts in Haulbowline and bought a lot of beautiful timber there. This timber came from ships being scrapped in the shipyard. Some of this timber was salvaged from the liner s.s. Celtic. He hired a local carpenter, named Cashman, from St. Colman's Square and paid him to make wardrobes and other bedroom furniture, which he then sold on. I loved the smell and fragrant aroma of the cut timber. He used some of this timber to upgrade the shop too. He put new compartments in the corners and built new partitions for the bathroom.

    I was lucky to have enjoyed every part of growing up and looked forward to many more happy events in the years to come, and also things that would help in my education and experience. I am sure there are youngsters who would love to have had experiences like these. The history-making events as well as beautiful sights in the town and Harbour are cherished memories I was so lucky to have seen and partaken in.

1940; Lynch kids, Jack, Eileen, Patricia Anthony ,Nora, Sheila & Mary

1946; Molly Kiely and Jack Lynch in Cork

Chapter nine

I attended the Presentation College in Cobh as teenager until the age of 15. Brother Aiden was the Principal there and it was he who taught me Latin. He was a quiet man and was known, to the boys at least, by his initials; WAC. Brother Regis, AKA Johnny Gob or Gobbo, was an older Brother and used to take us out into the adjoining field to learn 'about nature', as he put it, but in reality it was so that he could have a nap! English was his main forte. Out in the field he used to say; "Now be quiet, close your eyes, take in nature, and remember I'm watching you at all times." I don't think the poor man was up to teaching at his age.

I had a lay teacher we called Mosso, Moss O'Brien, who taught science, physics and mathematics. He was a brilliant teacher and also took the class for Rugby, which was controversial at the time as some Brothers wanted only Gaelic played in the College. The college rugby team did very well in college championship games though. A few of the lads got international caps. Mosso's favourite expression was, "Have you my knife?" which meant 'Do you understand me?'

Brother Liam, alias 'Billy the Kid', was young and aggressive. I had a run in with him once that I remember well. One day he was asked to keep an eye on us while our teacher was away. He taught his own class within the same room and at one point he gave a hiding to Tommy Donovan, a neighbour and friend of mine. When I looked across and nodded to Tommy to check he was alright, 'Billy the Kid' flew into a terrible rage. He came rushing towards me, his face red and the cane raised above his head. He shouted at me; "Get up," and I did. I was taller than him; he was only about twenty years old. Again, in a raised voice he shouted;

"Hold out your hand."

"Why?" I asked.

"You were looking across and disrupting my class and you were not studying." he shouted.

"I was looking in your direction only because I was learning my subject off by heart and I could not look at my book whilst I was doing this," I lied through my teeth, as I denied his accusation.

"Hold out your hand", he shouted.

I refused and said no, and then added "don't try and use that cane on me."

By now the two classes were completely absorbed with the confrontation taking place in front of them and I was actually enjoying it because I felt in control, for the first time. I was cool and composed, while he was frustrated and upset. Billy the Kid then lifted the cane and took a swipe at me. The cane hit me across the chest and broke a pen I had in my breast pocket.

"Look at what you've done," I said and I grabbed his wrist and took the cane from him. He was furious. His face was so red I thought he would blow a vessel.

He screamed; "Give that cane back to me."

"You can have it if you promise to take it back to your own class and do not try and use it on me again." I told him. He promised and I gave him the cane. Then he said; "Leave the class and go home."

I immediately packed my books and left the class. Soon after, my friend Cecil James came out with his books under his coat and we both enjoyed the afternoon together. The whole affair was never mentioned again, even when Billy the Kid took over our class the following year.

At weekends, or sometimes after school hours, daddy never seemed to be happy unless he had me working at something or other, particularly if it meant he did not have to perform the tasks himself. To keep me out of harm's way he decided to fit awnings over the shop and I came in handy. It was my job to get up on the ladder and punch four holes in the concrete to hold metal eyelets, which held the supports for the awnings in place. I sweated and swore under my breath throughout. There were many occasions when the hammer slipped on the chisel and caught my knuckles. Next, I had to fit hooks across the front of the shop, to hook up the length of the awning. Eventually, the job was finished and the awnings were up, in full glory, to the front of the shop and the customers were protected from rain and extreme sunshine.

Throughout my school going years mammy gave shelter to various school pals of ours who wanted to go on the 'lang' (play truant). On one occasion my friend Chris Walsh decided he was not going to school that particular day and called into the house. He went to the stove where mammy kept her cigarette butts and helped himself to a dried up Woodbine butt. I, too, took these butts on occasion when I wanted a smoke but daddy did not find out for some time that I smoked.

Chris felt safe and happy to be inside our nice warm house protected from the freezing weather outside. Daddy was out at work so he was safe enough. However, the bad weather this particular day caused daddy to decide to come home early and Chris, knowing he would be in big trouble if discovered, just made it out into the backyard in time, climbing up on the outside loo to hide. The tank for the loo was on top of the roof and was frozen over but Chris had not noticed because he too was freezing. Daddy came into the kitchen and rushed straight to the outside loo as he was having difficulty holding on and seemed desperately in need of relief. Chris nearly had a heart attack when he heard daddy go into the loo and prayed that he would move out soon. No such luck! Daddy sat on the loo and, despite the cold, took his sweet time. Chris was shaking and freezing, he didn't have time to grab his coat on the way out and he could not move to stretch his stiff joints for fear of alerting daddy. Eventually, daddy pulled the chain and Chris breathed again only to suddenly realise that the ball cock was stuck solid in the frozen water. Daddy was mumbling to himself, wondering why the loo wouldn't flush and Chris thought that his time was up, sure it was only seconds before daddy would come up to see what the problem was. Chris immediately broke the ice, plunged his hand and forearm up to the elbow in the icy water in the tank, and released the ballcock. The water flowed as Chris breathed a sigh of relief and daddy, relieved the toilet was back working, went inside. After some time mammy came out to tell Chris that daddy had gone out again and the coast was clear. Chris could not get inside quickly enough to thaw out and dry his wet sleeve. Daddy often spoke about putting a fur covering on the toilet seat because it was so cold. It's just as well he didn't because our aim was not too good!

All through our childhood daddy made sure we had shoes, even if they did hurt our feet, and enough clothes, at a time when other children ran around barefoot, cold and hungry. I remember he also gave food and clothes to others less well off than we were too. As with most children we were not always too satisfied with the clothes we wore. During the war, when things were tough, he had overcoats made from blankets. These were dyed dark brown. I also had clothes made from daddy's old suits. Usually, the material was turned inside out and stitched up into a suit or trousers.

When I was 11 years old I was lucky enough to see Eamonn deValera arriving to reclaim the ports for the Irish. Spike Island, which was known as Fort Westmoreland, and Fort Camden, Fort Carlisle and Fort Templebreedy were all taken back when he was Prime Minister of Ireland. Irish and British soldiers were present when the Union Jack was lowered for the last time over Spike Island, and the Irish Tricolour was raised. It was an electric moment. The National Anthem was played and everybody clapped and cheered and hugged one another. A twenty-one gun salute was given. At that same time there were also many sad farewells to British military personnel, as they had become very attached to the locals during their time living with us.

A local boat, the Morsecock had ferried dignitaries and officials from the Deepwater Quay to Spike Island to witness and take part in the ceremonies surrounding the auspicious moment. Other Irish ports previously occupied up to then by British Troops were also taken over at the same time. After the initial celebrations, reality struck home in the town and things went downhill, particularly when World War II started. In Ireland this was known as The Emergency. The whole atmosphere of the town changed then.

British Royal Naval vessels were no longer seen in the Cork Harbour and the town suffered from the losses in business and revenue. Local women who had married British soldiers and sailors left town, emigrating with their husbands to England.

Ireland was now isolated during World War II. Little did we know how much Churchill, the new Prime Minister in Britain, would want the ports back when war broke out. If he had been in power, instead of Neville Chamberlain who was the British Prime Minister involved in the handover of the ports, deValera might not have got the ports back as easily as he did. Churchill was no friend of Ireland. There were rumours that the British were going to invade Ireland to get the ports back when war broke out, as they were vital for Britain and the Atlantic sea war. It was also rumoured that America had somehow quashed the proposal. There was a big Irish influence with the American/Irish groupings.

In town we had two cinemas, which miraculously survived the losses inflicted on the area by the war and were a great source of escapism during those dark days. The Coliseum, which was part of the Young Men's Society building, was run by the Moynihans. The second cinema, The Arch, was owned by Frenetts and was run by Johnny Morgan and John Hennessy, two local men.

My favourite movies were 'Cowboys and Indians' films. Most cowboy films were associated with Indians in those days. Tom Mix, Ken Maynard, Gene Autry, Charles Starret, Roy Rogers were the actors of my day. Other well-liked actors were Tim Holt, Jimmy Cagney, Andy Devine, Lloyd Nolan, George Raft, Charlie Chaplin, and Laurel and Hardy. The singing cowboys, Sons of the Pioneers, rated high with me also. Our Gang, The Three Stooges, and Marx Brothers were also in my list of 'must see.' I did not rate female actresses much then as I was a bit young for that kind of interest, except Marjorie Maine, who was not sexy and who always appeared in cowboy films. I also loved the Tarzan films. Bulldog Drummond was a favourite, with Leslie Howard in the lead role. When I grew up, I fell in love with the gorgeous, dark haired Hedy Lemarr.

There was nobody like Hedy. My best friend Danny Hunt fell for another actress, Joan Leslie, who had fair hair. We talked about these actresses as if we knew them. One of the highlights of the cinema was the weekly serial, like X-Men. The agent was always on the verge of being killed at the end of an episode and we would try to figure out how he could be saved next week. It was a long time to wait for the hero to outwit the enemy. It was the topic of boys' conversations all week long. It cost four pence to get into the cinema and our parents sat in the one-shilling seats. During the war we got in to some cinemas by collecting jam jars, these were accepted in lieu of money. All the films were early black and whites and some were grainy, and there were lots of break downs.

One night, I was in the Coliseum cinema in the shilling seats, as I was older at this stage, when the local butcher's daughter came in and sat between me and another fellow. After a few minutes, when the cartoon had started, the girl let out a scream. I jumped and there was general pandemonium around us. The lights came on and the fellow on the other side of the girl looked frozen to the seat. Everybody was concerned and the girl explained that the fellow, whom she knew, asked her if she would like a sweet and with her eyes glued to the screen she dipped her hand into the bag and felt something cold, and slimy. This frightened the life out of her and she had very disturbing thoughts about the contents. As it turned the poor fellow had two bags with him, one contained sweets and the other contained sausages! She had dipped her hand into the wrong bag. He could not get out quickly enough!

The Four Feathers, a film depicting the British army in action, was showing in the Coliseum one night when a few of us went to watch a movie. During one scene, when the British flag was being raised, a bottle of black ink hit the screen. An anti-British demonstration was in progress. The show stopped and the Guards arrived and arrested one fellow whose name I can't remember. It ruined the rest of the film for us, trying to watch the screen with ink smeared across it!

There were two shows on week nights - one from 7p.m. to 9p.m. and the second from 9 to 11p.m. On Sundays there was usually a matinee at 3p.m. and one evening performance. Immediately after each show we would all stand to attention whilst the sound system blared out The Soldier's Song, which was the English version of our National Anthem. With gusto we all shouted our own words to it;

*"Soldiers are we,
whose lives are pledged to Ireland.
Some have come,
from a land, beyond the sea...*

Later, the Atlantic Theatre was renamed the Atlantic Cinema. In the early thirties the Atlantic Cinema was originally called The Baths, where the sign said 'Hot and Cold, Fresh and Saltwater Baths.' It was known as 'Queenstown Seawater Baths and Recreation Co.' and it was here where steerage class passengers to the U.S. were deloused before they boarded the liner which was on route to Ellis Island. When Ellis Island later closed the Baths became defunct and the building was taken over by Mossie McDonnell, a local butcher from Harbour Row. He got rid of the baths amid much controversy and converted the building into a modern cinema, just prior to WW II. Whilst waiting for the cinema equipment to arrive from France, war was declared and the equipment was held up in France until the war ended. In the meantime, the cinema became a theatre and we saw performances from Michael Mac Liamoir, Hilton Edwards, Lord Longford, Anew McMaster, Jack Doyle and Movita. The local Argosy Players, which included Denis Harrington, Mossie McDonnell, Kitty Sexton, Kitty Fitzgerald, Jasper Wilson, Pat Murphy, Nollaig O'Brien and many more also performed here.

It was in this theatre that Jack Doyle and Movita had a big fight that I'm sure most of the onlookers would still remember to this day! The relationship had been very fiery with Movita, and her Mexican temper, often coming out on top. I was looking out our bedroom window one night, watching the crowds going into the Atlantic Theatre. After some time, a lot of noise could be heard and after a while Jack came out, followed by Movita, and a big row seemed to be in progress. Our windows were wide open and mammy, daddy and Eileen were also shamelessly looking on. Many of our neighbours were also absorbed with the confrontation taking place. We had a clear view from our house! From what I could hear poor Jack had his face cut with glass and tried to put a brave front on himself. However, Movita took off for the States soon after and that was the end of their marriage.

Jack had a reputation for womanising and drinking, which affected his career. However, he seemed to have been nice to the locals. My sister Sheila later told me that she and some of her friends were playing down in the baths when Jack came up to them and gave them some money so he obviously has a soft spot for some of the local children.

The theatre was again converted to a cinema when the equipment eventually arrived and was installed. It was then named The Ormond and later renamed The Tower. Here, Johnny Cashman was the caretaker and ticket collector. Poor Johnny's eyesight was not great and the boys used to pass him all kinds of paper, including Woodbine packets, as tickets. Whilst he tried to apprehend one boy, another used to sneak in behind his back. It was common to see Johnny going from row to row with his flashlight, looking for these boys who hid on the floor to escape detection. As this occurred during a film, people shouted at Johnny to put the light out. The poor man had a rough time.

In Cork, one night, Dan Hunt and I had been to the Savoy Cinema, and when we came out there was a large crowd at the bus stop. We were in danger of missing our last train to Cobh unless we got on the next bus. Fortunately, we had some Chinese crackers with us. I lit one of the crackers with a lighted cigarette and casually walked up towards the front of the queue and dropped the cracker about three rows from the front. I then stood back. In no time, the cracker exploded, and as all good crackers do, it scared the lives out of the people who were not expecting it. They scattered and Dan and I moved in at the head of the queue, expressing our horror at what had happened. We caught the last train home just fine and laughed to ourselves the whole way home!

We got up to plenty of capers as young lads. We had a very complicated version of what kids today call 'knick knacks' that we used to carry out at night. We used to tie a string to a knocker on one side of the road and then take it back to the next house on the other side and so on until we got about six houses tied up. We would then knock at the first house, hide and watch and wait. The first person would open their door and look around, which would cause the string to be pulled tight on the knocker of the door opposite, calling them to their empty doorway, and so on. We would watch fascinated, as people up and down the street all came out at night wondering what was going on and then, when they realised they'd been had, cursing whoever did it. They never did find out who it was and we carried on regardless.

Another gem of a prank was also carried out at night. We used to tie a piece of black thread between a knocker and a nearby post, or petrol pump. The thread was pitched at head-height. In those days, most men wore hats. As they came to the thread, the hat went flying and the doorknocker was activated. The owner came to the door only to find a man groping in the dark, looking for his hat. It was hilarious and we got great enjoyment out of it, until one night it backfired on us…

There was a convenient petrol pump right in front of Barry's door, at number 34, which made it an easy target. On the other side of our house was number 32, Mackey's Pub. One night Tommy Donovan and I were inside our house, at number 33, and we decided to tie a piece of black thread between the pump and Barry's door knocker. Mammy gave us some blue woollen thread which she was using for knitting. We blackened it with black shoe polish. This turned out to be our biggest mistake. Anyway, we looked out and when the coast was clear, we finished laying the trap.

Mammy, who knew what we were up to said,

"Tommy, make sure your father does not come out of Mackeys and get caught."

Guard Donovan, Tommy's dad, used to go to Mackey's for his jar on a regular basis.

"No, he's on duty tonight," Tommy said, reassuring her.

Time passed as we looked out our shop window, hoping to spring the trap. Some women, and men without hats, got through our trap without incident. Then, all of a sudden, Guard Donovan, with his Guard's cap in a jolly slanted tilt passed our window.

"Oh Jesus," said mammy, who was also looking out the window.

Guard Donovan got to the pump and just as planned, though not for him, off went his hat into the darkness. We heard the knocker hit the door at Barry's and Charlie Barry opened it and found Guard Donovan down on his hands and knees looking for his black cap in the dark, on the road.

Mr. Barry did not know what was the matter and thought Guard Donovan was drunk and had fallen down. We were close to wetting ourselves with laughter but the laughter didn't last long. Daddy appeared with some other men and saw the same scene. They all then got together and had a pow-wow. Daddy came into the house and looked at Tommy and me, but did not say anything to us. Instead, he said to mammy, "You have blue wool for knitting, don't you?"

"Yes," she said quietly.

The smiles were gone off our faces by now.

Daddy then turned to me and said, "Show me your hands."

Finally, it dawned on me. There was the evidence of black shoe polish on both my hands. Daddy laughed loudly and said he had never seen anything so funny as Guard Donovan clawing in the dark for his hat. He too thought Guard Donovan was drunk and had fallen over. We didn't get in trouble thankfully, mostly because daddy found the whole affair so funny but this ended our game so close to home for the time being. I reckon daddy should have been a policeman after that detective work.

Robbing (or slogging, as we called it) orchards was a common occurrence in Cobh and all the children knew the location of the best, and most accessible, orchards in the area. One night, some of us robbed an orchard out the High Road and were on our way back to town when two guards came racing along on bikes. They stopped when they saw us to see what we were up to. One of the lads saw that one of the guards was his father and there were also two sons of the local sergeant with us as well. These boys started to get nervous, getting caught by the guards and your father at the same time would have been even worse. The guards recognised us and asked us if we had seen any boys out the road as a phone call had been made telling the guards that lads were in an orchard, stealing apples. We said there were some fellows who had just passed us and who had ran down the short cut to the Lower Road, via the stationmaster's house. They thanked us for the information and peddled off in pursuit of the fictitious boys. This was a close shave but not the only one we had.

Around the town the guards busied themselves checking for lights on bikes, following up reports of orchard slogging, checking wireless licenses, pub closing hours and making sure pub visitations on Sundays - when only visitors were allowed in for a drink - were being obeyed. Usually, the locals walked to Ballymore, or Carrigaloe, for a Sunday drink. A visitor was defined as anybody who travelled more than three miles to the pub. This was a Sunday stroll for the lads. However some fellows stayed at their regular pubs and took the chance of getting away with not being caught in a guard raid. Pubs owners pulled blinds and imposed silence outside of hours, and even posted their look-outs near the pub. There was one particular pub, down The Mall in the Holy Ground, which I believe was never caught in a raid, despite open drinking being carried on outside it. The famous Holy Ground is a street British sailors and others frequented in earlier times. Lookouts were placed at the top of Cotterell's Row and outside the pub and on the Holy Ground to watch the approaches to the pub. This meant that nobody could get near it without being seen. One time the guards came along in a boat and tried to surprise the owners permitting open drinking take place, but had no luck and had to give up trying. When an attempted raid was in progress, the boys casually walked outside the pub, with their pints in their hands, and laughed at the frustrated guards, who could do nothing as the pint drinkers were not, technically, on the premises.

Entrapment was not an unheard of term in those days. The guards were ingenuous and had lots of ways and means of catching those deemed to be breaking the law. The guards used Belvelly Bridge as a trap point to catch culprits who cycled home at night with unlit bicycles. It was an ideal trap zone as it was quite narrow and the guards had a good view in all directions. One night a poor fellow was merrily cycling along with no lights on his bike, around eleven p.m., when the local Sergeant called on him to dismount. The fellow took fright, hearing the voice coming out of the dark from seemingly nowhere, and made a dash for freedom, only to find the Sergeant's great overcoat thrown over him like a giant net. After he got up, gingerly rubbing his injured leg and ego, he had his name taken and had to appear in court at a scheduled date. There was little opportunity for escape from that point.

During the Emergency the local guards found they had much more opportunity to deal with serious crime and their skills were really put to the test. When the Royal Cork Yacht Club came under attack from the IRA during that time and was threatened with being blown up guard Dunne came on the scene whilst on duty and was nearly shot by one of the gang. I heard that the bullet lodged in the wall quite close to him as he entered into the building to investigate the break-in. The gang of three, or four were later identified but nothing could be positively proved against them!

The Emergency affected every aspect of everyday life for us, even though it was far from the war and we were supposedly neutral. Due to fuel shortages the trains were very unreliable around the rest of the country. Nobody knew when a train was due to arrive or depart. The train drivers, who operated between Cobh and Cork, sometimes stopped the trains in Fota, where they would get out collecting timber in the woods to use as fuel to keep the trains moving. The drivers worked very hard and in those lean times there was a lot of pitching in to be done.

In the home the shortages were very noticeable; tealeaves were brewed over and over again to extract the last drop of flavour and colour from them before they were then dried and re-used as a tobacco alternative. Porridge was the mainstay of the breakfast and gone were the days of cooked breakfasts except on special occasions. Usually, there was a lot of chaff in the oats, which got stuck in the roof of my mouth, or in my throat. I hated lumpy porridge and daddy loved it, so you can guess which kind I got most of the time.

When rationing was introduced, everybody was issued with a book of coupons. These coupons itemised the allocation of foodstuffs due to each person. Tea, sugar, butter and various other goods to be purchased with the coupons had to be weighed in the shop, as nothing was prepacked.

Each household was allowed certain rations of critical items per person, per week. The larger the family, the better off they were. We mostly collected our rations from Mr. O'Kane, who had a grocery shop on Harbour Row. When I went for our ration of butter each week he would take out a lump of butter and pat it into shape using butter bats, until each lump weighed a pound, or half pound, as I wanted. He really took pride in this operation and he kept the butter in a consistent rectangular shape. He would then remove the coupons as necessary from the book. Other shops just weighed items and did not take too much trouble with presentation. Farmers delivered their country butter in ten pound lumps and this had to be patted in the same way as the creamery butter. It was usually much saltier than creamery butter and had a much more yellowish colour.

The Weights and Measures guard used to go around regularly checking weights, and measures, in the shops to ensure they were correct so that the rations system was not being abused. There were rumours that some shopkeepers doctored the scales and gave short measures, though I can't say that I knew of any such cases.

I always remember that mammy loved her cigarettes and when the war started, with cigarette and tobacco supplies very scarce, she suffered with the lack of them. During the Emergency cigarettes and tobacco were kept under the counters in shops and doled out to special customers, such was their scarcity. Mammy's supply was not meeting her needs so we had to keep our ears to the ground and listen carefully for whispers and rumours of which shops might be hoarding some supplies and then we scooted to the shop and begged for some for mammy. Back then there was no age limit on children buying cigarettes for their parents.

Tobacco and the lack of it was a common topic of conversation. People discussed how they were getting on with their addiction and shared information on what they were doing to feed the habit. These included smoking dried tea leaves, dandelion roots, rat leaves. Rat leaves, as we knew them then, were large leaves on long stems that grew from the ground near walls and in damp places. Nothing took the place of the nicotine though; all old butts were recycled and re-rolled over and over again to get the very last possible drop of nicotine from them

The only time the ration system gave way a bit during the Emergency was around Christmas when our grocers gave out Christmas boxes to their customers. We normally shopped at May O' Sullivan's and at O' Kane's, so we usually got two Christmas boxes. Sometimes the box contained a Christmas cake or a ham; we might get a box of chocolates too. It was something we looked forward to but it all ceased when war broke out.

Of course the other thing that rationing also brought to the fore, alongside an ability to make do, was an increased lack of hygiene when it came to food. As food became scarcer and scarcer people's standards of hygiene dropped and as needs grew, our parents' tolerance for the unhygienic increased. The shop-keepers around the town used to place the vegetables on the pavements outside the shops when they were first delivered, and dogs would come along and urinate all over them and nothing was cleaned before sale. It was very different to how it is now. People never washed their hands after handling the soiled vegetables and flies were everywhere on the cooked and uncooked foods. I think we must have had every germ and virus going without being aware of it. There was no such thing as keeping cooked and uncooked meats apart and there weren't any fridges or freezers to store the food in. There were no detergents to clean and degrease surfaces and household utensils. During the Emergency all these issues became even more problematic now that everything was so scarce. Brown paper, used to tie up the food before a purchase left the store was becoming less and less available and so often food was just carried out as is.

Home grown produce, like milk, was often easier to obtain at this time as it wasn't brought in from outside. Milk was delivered to houses directly from the farm then. Paddy Barry used to come along in his pony and trap and measure out the milk from metal churns. The measuring device was called a 'Pawnee', which was a tin half pint or pint size container on a long metal handle to reach down into the churn. Usually we got a little 'sup' for the cat from Paddy when he arrived. He was always very nice to us and we were pleased when he later married my Aunt Kate. There was another milk supplier in the area that some people used, though my parents never did because the milk was warm and watery and sometimes had hairs floating in it. At that time all the milk came straight from the cows and went direct to our doors so I don't think there were any health checks on the animals.

We were very resourceful in many ways during those lean times. It was impossible to get petrol products and coal so turf and wood became the main heating fuels. Daddy hired a plot of ground in Carrignafoy and we grew our own spuds, cabbage and various other vegetables there. I used to have to dig trenches, sow the seeds, and cover the vegetables with manure. I was the one who had to do the digging and bring the vegetables home and yet everybody ate the fruits of my labour! I caught plenty of fish for us to supplement our meals and preserved them by salting or smoking them. One summer daddy decided to build a fish-smoking unit in the backyard. The basic equipment was a cleaned out five-gallon oil drum with the lid removed. A hole was cut out in the centre of the bottom of the drum. A broom handle of the same diameter was now inserted into the hole and the sawdust was packed tightly into the drum. The broom handle was then removed and the drum was placed on a few bricks where some kindling was burned under it. The flames licked into the hole and the sawdust continued to smoulder and give off smoke. The smoke soon filled the inside of the purpose built enclosure, which had lines strung from side to side. We hung mackerel, congers, whiting and pollock on the lines until they were smoked. Over-smoking caused the fish to be virtually cooked, in which case we just ate them straight from the smoker, or from the clotheslines in the kitchen where the smoked fish were hung for use later.

It was one of my jobs to get bags of sawdust for the smoker from a builder named J.J. Healy at Top O' The Hill. I used to put a full bag of sawdust in between the crossbar and pedals of the bike and freewheel down Middleton Street, around the cathedral, and down Harbour Hill back home. I recall one trip that ended in tears; I was coming down Harbour Hill on the bike, which was loaded with a big bag of sawdust. Just as I got to Harbour Row a young boy ran across the path in front of me, I immediately hit the brakes but it was too late and I hit the ground almost at the same time. I came over the handlebars and remember nothing more except May O'Sullivan, the shopkeeper from nearby, asking me if I was OK.

It was three days later when I awoke in bed, with mammy and daddy leaning over me, that I first spoke;
"Was it love?" I kept asking them.
They said I had concussion. They told me that I had walked home without assistance and came in with my forehead bleeding and looking completely dazed. I remember nothing about the walk home. The doctor visited the house daily over the next few days to keep an eye on me. Days later, when I was up and about, they asked me what was I talking about when I had asked them 'was it love?' and strangely I had a recollection that the boy who ran in front of me was Clayton Love's son or nephew. Clayton

Love had a shop at the top of East Beach, close to where the accident happened. I was assured I had not hit him and that he was fine. I do not remember getting any more sawdust after that, not because of loss of memory, but because daddy would not let do it again after my accident

    Food was so precious it was rare to let it go to waste but there was an occasion I remember when I foolishly did just that. I visited Aunty Lina, in Ballymore, when I was about fifteen years old with a few of my friends and we got a warm reception from her with some delicious cake to eat and cups of tea. When we were leaving, my aunt gave us a supply of potatoes, a dozen eggs and a chicken to bring home to my parents. We cut through the woods at Cuskinny on the way home to make the journey a bit shorter but I was finding carrying the eggs a burden so we divided into two groups and each group got half the eggs to carry. However, being teenage boys our high spirits got the better of us and we proceeded to have an egg fight in the woods. Eggs flew in all directions and we looked a sorry mess when we got home. I sneaked into my room washed myself, changed, and put my clothes with the next load of washing to try to hide the awful waste of food, something I would surely have gotten in big trouble for, especially with rationing in force.

    Daddy was inventive when it came to making do and saving money during the days when things were tight. I remember he tried to re-sole damaged boots with tyre rubber. When he found it was not a success he used leather and I later also became quite good at heeling and resoling shoes. I got cuts of leather from O'Donovan's shop, and soaked the leather in water to soften it before re-shaping the piece to suit the shoe, or boot.

    Boys wore boots with metal studs, to save wear and tear and the boots were not always the most comfortable and often hurt the toes. I don't remember any soft leather being available. We ran and skidded on the studs to leave a shower of sparks as metal and stone touched.

    My brother Anthony was lucky as he was too far behind me in years to have to wear my discarded clothes or shoes. It was worse for my sisters as they had to put up with hand-me-downs irrespective of size and shape. Being the eldest had its advantages.

Around this time daddy decided he was going to do a job on the kitchen ceiling. The ceiling was cracked and had been whitewashed but was peeling and cracking again and needed something done to it. With the aid of a friend, Jimmy Sheehan, he decided to use panels of plywood and beading to cover the ceiling. At one stage Jimmy was up on the ladder, above daddy, and dropped the hammer on his head. A large lump quickly appeared on daddy's head.

"Blast you Jimmy, will you watch what you're doing?" shouted daddy.

Poor Jimmy was all apologies and kept muttering to himself.

Daddy then painted the ceiling a dark blue and since he knew in advance did not have enough paint to finish the job he added plenty of linseed oil to eke it out. The paint never dried and was patchy and sticky for as long as I can remember.

Daddy also used to put paper on walls, where it belongs, and on, where it doesn't. To understand the trauma of this work it is necessary to realise the problems associated with it. In those days, paper came in rolls, with borders attached, and these had to be trimmed straight by hand. The borders were about a half inch wide and only identified the wallpaper manufacturer. Most rolls were patterned and there was no vinyl available. We tried to keep a straight line, with a bad scissors, and as we unrolled the paper, it often tore, causing a lot of swearing. When we got the paper cut we now re-rolled it, and measured it for ceiling to floor length. We had to make sure the pattern matched. Now the paste, had to be mixed and this was ordinary flour and water. When the paste was applied to the paper, more swearing as the soggy paper ripped when it was lifted. As far as I remember, only paper manufactured in Kildare, for some reason, was available. That part was bad, but worse was to come. As I said, the patterns had to be matched and the paper edges had to be aligned. It usually ended with overlapping paper edges, misalignment of patterns, and various residue patches of dried flour all over the paper. And that was only the wallpapering!

The border paper was a finishing touch which helped conceal the defects between the ceiling and wall.

The nightmare really started when daddy wanted to paper the ceiling. After the preparation outlined above, he climbed up the ladder and started to align the edge of the paper with the junction of the ceiling and wall. For ease of handling and controlling the paper was folded like a concertina. It was my job to stand underneath with a floor brush and hold the folded paper in position, while he moved along the ceiling and patted the paper in position. My arms ached and the brush got heavier and heavier, so that I could not hold the paper in position, and soon the paper started to fall off behind me as daddy moved forwards. Eventually, we finished the job, which had the same results and appearance as the walls that had been papered, though perhaps there were even more errors and defects, if possible!

Daddy, in his wisdom, came home one day, with a large tin of paint. Joyfully, he said it was to paint the lower half of the front of the house. He started the job, with me as the eldest helping him as usual. As we layered on the reddish brown paint we realised it looked terrible. Someone even remarked that they thought daddy had painted the house with marine anti-fouling paint that daddy had got cheap in the Haulbowline shipyard. It was a complete disaster. We had to start all over again and repaint it with 'Blessed Virgin' blue and white. I felt like praying every time I looked at it but daddy thought it was great!

In 1945 the war finally ceased and despite all the obstacles our family had survived without any great hardship. All credit goes to daddy and mammy for taking care of a family of two boys, and six girls during those tough times.

Chapter ten

The Power House was a red brick building with a large chimney stack that was situated in the Bath's Quay. It was used to heat the baths used by the emigrants going to the USA. The stack is now the only part left. I was about seventeen years old when I started to work in the decommissioned Power House stacking turf. This was hard and dirty work and I was eaten alive by fleas, which thrived in the turf. My fingertips were tender and sore from the work but as time went by they hardened and gradually came to resemble leather. During the war, turf was brought straight from the bogs and a lot of it had not been dried out.

It annoyed me when I saw the cruel treatment inflicted on the horses used to draw the turf. The turf used to come in at the railway station or to the Deepwater Quay and was loaded into carts and the horse owners were paid for each load which was then taken to the Power House. The more turf they delivered the more money they got. Therefore, the horses were driven as fast as possible, in each direction. In winter the poor horses used to occasionally slip on the icy roads and they had great difficulty getting up with cut knees, which were not treated. These runs went on all day until dusk, and the poor horses were knackered. One particular owner used to beat the poor horse across the head, to make him go faster. I saw blood running from the horse's head. This man had a bad temper and his cruelty was unbelievable. There were no organisations to prevent cruelty to animals then and I didn't witness the Guards or any concerned people take action to prevent the man from inflicting hurt to the poor horse.

I also did relief work for the Insurance agents whilst they were on holidays to earn some money. I got to know a lot of people this way and they trusted me to enter their houses whilst they were away. People were much more trusting then. I knew where they left the premium money and book and would take care of my business and leave the house as I had found it.

Some cases I came across doing this work were heart rending and really touched me. There were old ladies who could barely afford to pay one penny per week. One particular house always stays in my memory. I used to climb three flights of stairs to the top of the tenement to collect from a woman who was always in her room but whom I never saw during the years I collected from her. As I got up towards the third floor, the smell used to make me heave. There was no light on the landing, which was pitch black, and the room was in total darkness. The smell was overwhelming and many times I felt I would get sick. I used to take a deep breath lower down the stairs, rush up, collect the money from a known spot, mark the book and belt it downstairs as fast as I could. Other lonely people loved to chat to me when I came to collect and I liked being able to bring a bit of happiness to their day.

  Working with daddy could usually be easy-going but I do remember once having a big row with him over our shared responsibilities now that I was collecting insurance money. He had asked me to deliver an Insurance renewal form to a house and I had forgotten to deliver it. Next morning we were all at the breakfast table eating porridge. He asked me if I had delivered the form and I said I had forgotten but that I would do it that day. He asked me for the form, looked at it, and then threw it back at me, telling me to make sure it was delivered that day. His face was red and he was in a real temper. Something had set him off and I could not understand what it might have been. It was not my failure to deliver the form - that was for sure. The form landed in my porridge and without meaning to I threw it back in his direction. He immediately got up with a pot and came towards me in a threatening manner to hit me. I jumped up, picked up a small wooden stool close by and held it over my head for protection. Mammy was petrified and Anthony and my sisters gaped in amazement at the scene playing out before them. I looked at my father and said, "If you come at me with that pot, I won't be responsible for what happens."

  There was dead silence as my father stepped closer to me. I threw the stool on the tiled floor between us, where it broke into bits. Daddy stopped dead in his tracks, went pale, and left the kitchen without a word. I said nothing, took the form, delivered it, and came home later after much contemplation about the events. I justified my actions due to my age - I was seventeen- and because I felt it could have been more serious if he had hit me with the pot. As it was, only pride was hurt. He came home, that evening and I went up to him, apologised, and said I was sorry for what happened. He smiled and said he was also sorry, and we both hugged. He said he had not realised I was now grown up, with a will and temper of my own. We never again had a difference that was not resolved by talking and in a way that was a real turning point for us.

Around this time daddy converted and opened the office as a confectionery shop. Nicholas O'Keeffe from our road was a confectionery wholesaler and he supplied the sweets and other goodies to daddy. We went to Barrack Street in Cork to get supplies of hard-boiled sweets for the shop that Nicholas didn't supply and I remember being thrilled watching the sweets being manufactured. A lot of these hard boiled sweets are still available today but some of the more unusual which we bought were Acid Drops, which were square in shape and had a very bitter taste, and Money Balls which were like a small ball, reddish in colour and contained various shiny coins. Many children damaged their teeth biting into these sweets. With today's regulations there is no way these dangerous type sweets could or would be sold. A popular chocolate bar, which I have not seen for many years, was called 'Half- Time Johnny', identified with a wrapper showing a young hurler. 'Captain Mac' too was a very popular bar of dairy milk chocolate.

Mammy used the shop to supply her needs for cigarettes and was delighted with the new business venture. I loved the cider we sold, which was classed as non-alcoholic and was really apple juice, so I too was guilty of not paying. The cider was sold by the glass from a gallon jar. The whole family took what they wanted without paying for it. We did not realise that cash flow was important and just helped ourselves. Mammy could not say no to anybody and gave out a lot of credit too to friends and neighbours. Daddy had his own favourites too whom he allowed credit and as credit was not always repaid our debts mounted up and finally, the confectionary shop closed.

Years later, daddy re-opened the shop and sold women's and children's clothes. He did not have a clue about women's fashions so my late sister Mary ran the shop for a while. Mary wanted to buy in clothes for the younger generation, but daddy wanted to keep the older generation coming in to the shop so there was conflict over which avenue to pursue. As it was that venture ended after a short while too. This was again mainly due to daddy interfering in a product of which he had no knowledge!

Even as I got older I still had a fondness for practical jokes. One time during an election campaign in the Square, Martin Corry, a Fianna Fail candidate, was addressing a big crowd and Dan Hunt and I were watching what was going on with great but non-political interest. I had a mechanical toy mouse, which was very realistic, with me and I wound it up and waited until we saw a group of women at the meeting before letting it off. In the middle of Martin Corry's speech I watched the mouse scuttle towards the women and hit one on the shoe. She looked down and screamed; "mouse, mouse, mouse!" Her friends panicked. All hell broke loose as people ran, shouted, and generally lost their heads. We doubled up laughing and even though I lost my toy mouse I felt it was worth it to see such havoc caused by my own hand!

Prior to the Emergency, the British Army, Navy, and Irish civilians frequented the Soldier's Home, a meeting place for the soldiers where meals were available whilst they were in the area. Civilians could also use the facilities available there. After the Ports were taken over, it became a haunt for Irish Military personnel and civilians. We children and teenagers used to go in there regularly to buy Chester Cakes, which were our favourite, and were sold there in the canteen.

The home was run by two middle aged fellows from the area. One was known as Lofty. Unsurprisingly, he was very tall and thin, and had a high-pitched voice. The other was short and his name was Bertie. He too had a squeaky voice. Anthony, my brother, used to go in there with his mates and they used to go up to Lofty and ask, "How much are the cakes up there, Lofty?" They would point to the highest shelf, whilst keeping an eye on the Chester Cakes on the lower shelf. Lofty would turn away and strain to see the price and as he did so the boys would fill their pockets with the cakes. When Lofty turned back they would shake their heads and say the cakes were too dear and walk outside to devour their ill gotten goods.

One evening, I was with a crowd of lads when we went into the home to see what kind of mischief we could get up to. As we entered the home, we passed the mains electrical fuse box, which was at arm height. We got a few Chester cakes and on the way out I put my hand up and unscrewed a fuse. It was plain luck that the fuse I undid was the one for the main hall and shop. Darkness immediately descended, amid the clatter of knives and forks on plates, accompanied by raised voices. We got out as fast as we could and went across the road to watch the developments unfold. People started to come out and little flickers of light appeared inside the building as candles were lit. We decided to go up to my house and come back later. When we got there daddy met us at the door and said, "Good, you're back. Get your friends to come in too." I didn't smell a rat until I saw all my sisters, and mammy, smiling as we came in. In the kitchen all the chairs were spread around in a semi circle and rosary beads and prayer books were laid out on each one. Suddenly, it dawned on me. It was the First Friday and we were due for the Holy Hour in the house. There was no way I could get away now to go back and replace the fuse in the soldiers' home. I felt really terrible. The holy hour did me no good at all. All I could think of for the entire hour was the Soldiers' Home, in darkness and all the inhabitants wondering what had happened. Had they seen me loosen the fuse, and would the guards be calling here in the middle of the holy hour to take me away? I couldn't relax for a second. Rosary, after Rosary passed, Litany after Litany, Prayers for the dead. Prayers for the living, prayers for the missions, prayers for this and prayers for that, I thought the hour would never end. Mammy and the girls were giggling to themselves and daddy kept telling them to be quiet.

After what felt like hours, we got up off our knees and headed straight back to the home. We decided to disguise ourselves before entering just in case. Dan and I changed jackets and so did some of the others. Idiots! It did not enter our tiny minds that we were well known in the home and that even if we had worn skirts and blouses we would be instantly recognisable! Anyway, we went in and found Lofty and his mate working away in the shop by candlelight. Customers had candles on the tables and all seemed fine. I asked Lofty what the problem was and naturally he explained that they lights were gone they were waiting for the ESB. It had been over an hour since they had asked for assistance, he told me. I volunteered to have a look at the fuse box for him and he was very grateful. All the boys gathered round me to cover what I was doing and I took a few minutes before I reset the fuse to avoid suspicion. The lights came on again and there was great applause from all in the hall, as candles were extinguished. Lofty gave Chester cakes to all of us in thanks. We accepted with feigned reluctance and retreated to the sea front to enjoy our repast. Perhaps the Holy Hour did us some good after all…

    By now it was coming up to Christmas and carol singing was in full swing in the town. We decided to do it for the first time ourselves. Slightly embarrassed, we decided to go outside town and we headed for Monkstown, the next village across the river. We were doing fine singing and making a few pence until we came to one house. We had just started our rendition of 'Holy Night' and a very cross man came to the door and told us to go, as our shouting was waking the baby. He gave us a penny and we carried on singing from door to door… I cannot remember how much we got but we were delighted with ourselves. One of the carols we sang was;

See amid the winter's snow,
Born for us on earth below,
See the tender Lamb appear,
Christ was born on Christmas day

It took us some time to realise the third line was not, "See the tender land at the pier." We claimed we were singing for a charity and felt obliged to give some of the money to that charity, after we had deducted our expenses of course!

After the Government took over the Ports in 1939 one of the naval ships, the Muircu - which translated means Hound of the Sea - flying the Tricolour, having previously flown the British flag when she was then known as Helga. She had shelled the Four Courts in Dublin during the civil war. The Muircu and the Fort Ranock were taken over by the Marine Service and were stationed in Haulbowline. The Fort Ranock was used as a minelayer during the war, and later for fisheries patrols, whilst the "Muircu" was used for both coastal and fisheries patrol. Both flew the Tricolour and were originally fishing vessels that were converted to naval support. The Marine Service also used the salvage ship, s.s Shark for mine clearing. Later these three ships formed part of the new Naval Service and were the forerunners of purpose built vessels for the Naval Service. During the war Cobh residents got used to the sights and sounds of the Torpedo Boats roaring in and out of the harbour. They had been purchased from Britain. Originally there were six, named M1 to M6, but these dwindled as, one after another, they were cannibalised for spare parts to keep the others in operation. The Flower class corvettes Macha, Maeve, and Cliona followed and were based in Haulbowline which is still the main Irish Naval Base. Members of the Slua Mhuire trained on these boats - the Slua Mhuire was the naval equivalent of the FCA.

Visitors and emigrants no longer visited the town now that war had well and truly broken out and we missed all the excitement and buzz. The salvage ship, Shark, continued to come into harbour with salvaged parts from the liner s.s Celtic hanging from a lifting crane at her bow. The Celtic was not the only ship to founder off the mouth of the harbour. This was in December 1928, but salvage work continued for many years. Some ships, which arrived in the harbour during the Emergency, had anti-magnetic strips around their hulls, to counter the threat of the magnetic mines in the sea which would be attracted to the steel hull of the ships. Irish ships had the tri-colour blazoned across both sides of the ships' hulls to clearly identify them as neutral during the war. All of these ships were named after Trees; Fir, Larch, Beech, Popular, Willow, Oak, Rose, Cedar etc. Some had been impounded and renamed due to wartime regulations about overstaying in a neutral port whilst their country was involved in the war. At least one was found abandoned in the high seas, was towed to Cobh, refitted and renamed.

During the Emergency, the engineering corps prepared all the quays and piers for demolition in case the need arose to protect our borders and entry points. I watched them as they bored and drilled holes at regular intervals along the length of the quays. The holes were then covered with large concrete slabs that had metal rings for easy removal, when necessary. If, and when required, explosives could be placed in these holes to blow up the quays.

The Eighth Cycle Corps were a regular sight in the town and they were stationed up East Hill, at the Battery barracks. These were mainly young kids, who joined the army during the Emergency. We called them the Eighth Army. The locals felt that these young lads were harshly treated.

May Sunday, the first Sunday in May, was a day we all looked forward to with great anticipation. From the first time we heard of it we never missed a year without this being one of our prime days out. May Sunday was a festive day out in Glenbower Woods, Killeagh, about fifteen miles from Cobh. On this day, every year, visitors came from all over the city and county, and assembled for picnics. There were various groups drinking, singing, and looking for dates with the opposite sex. The latter was our chief motivation.

Our first trip there was on my friend Walter Barry's sidecar. There was Dan Hunt, Mickey Hunt, myself and four others, including Paddy Barry, who was in charge of the nag. When I say nag, I'm being kind. This horse was well past his day for glue making. The poor horse was close to being issued with an old age pension. Anyway, early on the Sunday we set off for Glenbower and passed through Belvelly, Carrigtowhill,

Midleton, and into Killeagh. By the time we reached the entrance to Glenbower woods the horse was dragging and sorely in need of a rest. We were worried about what would he be like on the way back, having to climb Ballard Hill and the other smaller inclines on the route home. Our throats were hoarse from singing:

*"We don't know where we're going until we're there,
There are lots, and lots, of rumours in the air,
We heard the captain say,
We're on the move today',
We only hope the blinking sergeant major knows the way."*

It was our first trip there and we were not disappointed. We had a great day but were outclassed by the Cork lads, who seemed to have a better technique at pulling the girls. Also, they had transport to offer the girls, such as bicycles with crossbars, or spare capacity in charabancs. Our only enticement for the girls was a half dead nag and no spare seating. We vowed that the next time would be different and we would come by bicycle.

When it came time to go Paddy hitched the rested nag up and we climbed aboard. When we started on the way back it was late evening and everything was great until we got to Midleton and found that Paddy Barry had no light for the sidecar and night had closed in. We were worried about the guards so Paddy made the horse gallop through the town really quickly to avoid being caught without a light. This was a big mistake. The poor horse was shattered and once outside the town it couldn't go on. We had to walk and push the sidecar up every hill back to Cobh. Like the horse too, we were all knackered and this reinforced our promise to take bikes the next time.

Year after year, the bikes came out and the numbers joining us grew on our big excursion. We had no luck with getting girls to ride on the crossbars though. One trip, when we were returning via Fota we came to The Wolf Lodge, which was rumoured to be haunted. Apparently the sculptor who carved the Wolves on the gates allegedly committed suicide and his ghost haunts the place since. Supposedly, the reason for his suicide was that he had forgotten to sculpture the tongues into the Wolves - his error shamed him so much that he took his life.

It was night-time, nearly midnight, pitch black and there were no lights on the roads or on the bikes as we approached the gates of the lodge. Dan Hunt, Mickey and I were in the lead, with my brother Anthony, Chris Walsh, Mike Walsh, Jimmy Donovan and Denis Ellis some bit behind us. Fear was in the air and as Dan and I went past the gates of the lodge we shouted, "GHOST, GHOST," and we peddled as fast as we could. Behind us we heard screams and suddenly there was an almighty bang, as unlit bikes crashed into each other and panic reigned in the tangled mess. Dan and I went back to see what was happening and could not believe our eyes. One lad was on his knees in the middle of the road reciting the Hail Mary and promising he would always be good from now on. That was one big fib! The others too were scared and so were we but we could not let them know that at the time. Eventually, bunched together, we navigated the road, past the lodge and no more mishaps. We laughed a lot about this as years went by.

The horse drawn charabancs that assembled in the woods were usually full of 'Shawlees'- Ladies from the Coal Quay - from the city and they had their stocks of crubeens, also known as pigs' trotters for sale. Bread, spuds and drink could be found everywhere. The smell of crubeens and stout was strong. Children ran around screaming and having a ball. It was usually glorious weather and this particular day was no exception. To top it all off we could not believe our eyes when we saw all the lovely girls going around in groups, like ourselves, on the lookout for dates.

On one visit, Dan and I were doing our rounds in the woods when we spotted two lovely girls walking together. One was blonde and the other was auburn haired. We immediately tagged on to them and got chatting and we both naturally decided which one we fancied, without even discussing the possibility that they would not fancy us in return. The blonde was my choice and her name was Theresa Mannix. I cannot say why I should have chosen her but we got on very well and promised to meet in Cork the following week. Dan came along and met up with her friend. We went to the pictures and had a great time and Theresa came down to Cobh with her friend the following Sunday. I got a photo of her when she was here and we fixed a date for the following Saturday in Cork.

During the following week my Aunty Lina came into our house and the photo of Theresa was on the kitchen table. Mammy had seen the photo but had not made any comments about Theresa.

Aunty Lina asked; "Where did you get that photo of Theresa Mannix?"

"Do you know her?" I asked.

"Of course I do, she's your third cousin." Aunty Lina replied.

I was dumbfounded. Now my romance was over. My third cousin! Surely I could not keep going out with her now? I told Theresa what happened and said I could not go out with her again. Now I know there was no reason why I should not have kept contact but for some reason then I thought it was terrible to go out with your third cousin! Much to my regret I never again saw Theresa. Even my mother did not have inkling about this cousin. Later, when I started tracing the Family Tree I tried to locate her family, without definite success. I did find out in the 1901 census that my mother was at the house of Jeremiah Mannix, near Kanturk when she was 7 years old. Jeremiah's wife was O'Regan.

This untimely end to my first fledgling relationship did not end our trips to Glenbower thankfully. On one occasion we cycled on to Garryvoe strand, which was not too far away, for a picnic. We decided to light a fire and fry some sausages. We located a wall about three feet high on one side and about five feet on the far side. Since a number of people were sitting behind the wall facing the sea, we decided to build the fire on the other sheltered side of the wall. We collected paper and kindling and Dan and I lit the fire. It began fine but gradually, as we poked at it, a large cloud of black smoke hit us between the eyes. Choking, coughing, and with eyes burning we jumped up for fresh air. It could only happen to us. There on view, in front of our faces, not more than one foot away at the other side of the wall, was a large bare female bottom. Dan and I saw it at the same time and were shocked to say the least! Peals of laughter came from the nearby people who witnessed this poor unfortunate woman trying to pull up her bathing suit, while we had her dead in sight. Needless to say we beat a hasty retreat.

Another time we were leaving Glenbower with two Cork girls that we picked up when my bicycle chain broke. We couldn't find a piece of rope for one to tow the other but the girls came to the rescue. Both had coats with belts and they offered us the belts which we gratefully received. We tied both belts together, and each end to each bike and took turns cycling and towing. All the way home we were praising these generous friendly girls and swore we would never forget them, and their kindness. I still remember their generosity, and the colours of their coats, but have not the foggiest idea of their names.

Walking up and down Patrick Street in Cork, affectionately known as Panna, was the well-known courting ritual of my day. Panna was a great place back then. There were some lovely girls who strutted their stuff along the street that I remember. We always got a date. Usually, we went to the Arcadia dance hall and it was here that I first learned to dance. I remember one night when Anthony asked a girl to dance and got the following reply: "Ask me friend, I'm sweating boy." He told me he asked another girl, "If I asked you to dance with me would you?" and when she replied "No" he gleefully answered; "Well I didn't ask you, did I?"

Dan and I used to watch the talent on the floor before we made our move for particular girls that we had our eye on. Sometimes we were lucky and other times we left not having kissed a girl at all. Usually, when we hit the jackpot we walked the girl home and most likely they lived in Blackpool or up Military Hill which was wonderful as it took a fairly long time to get to their houses, with long stops along the way. It was all innocent fun and a kiss was a great result.

With no dance scheduled on one of our usual week evening strolls Dan and I were at the Deepwater Quay watching Joe Murphy our friend, steering his motor launch filled with workmen in towards the Deepwater Quay steps. This was unusual because the launch usually left from and returned to the Camber. As we watched we heard shouting from the launch and saw that the launch was getting lower and lower in the water. We could tell, then, that something was wrong and the launch was leaking water and starting to sink. There were quite a number of people around and we all rushed to help. When the launch was almost six feet from the steps it disappeared completely into about ten feet of water. Fortunately, someone had grabbed a rope and all the workmen got to the steps safely, even if wringing wet. The funniest part of all was that Joe Murphy was Skipper of the boat, but was not the last man to leave the sinking ship. As soon as he got the chance, Joe was ashore. When the boat was submerged all we could see was the tip of its mast above the water. Serious and all as it was, we could only laugh and take the Mickey out of Joe for his swift desertion.

Living by the coast it was a common enough occurrence to witness, or be part of the odd sea-scrape. One that particularly stands out in my mind is a sea escape that happened to my brother Anthony, his friend Mike Walsh and some others. They had gone out in a motor launch to Rocky Island and pulled in to look around. One of the boys, Chris, picked up a large piece of wood and threw it into the boat, to take home for firewood. Unfortunately, it went right through the boat, creating a large hole. The boys tried carrying out repairs with no success and with night approaching nobody liked the idea of cohabitating on the island along with its furry inhabitants, large rats, which were living and breeding there. They had been told that these rats were a cross between rats and rabbits, some of the rats were supposed to be endowed with long furry tails, and this terrified the boys. Their bravado waivered and eventually they were frightened enough to realise they were going to have to find some way to attract attention.

It was a difficult spot to attract attention so they shouted with all their might but to no avail. Rocky Island had been used by the British as an ammunition storage depot during their occupation and had long since been abandoned. There was no one around and it was far from the mainland so their shouts weren't heard. Suddenly someone had a brainwave. One of the boys took off his shirt, doused it with siphoned petrol from the engine, attached it to an oar and set it alight. This torch was then waved backwards and forwards in a bid to catch the attention of someone. Fortunately for them, the Spike launch was on its way back to Spike from Cobh and diverted from its course to check on the flames seen by the crew. They picked up the lads, who were now very frightened but relieved, and took them to the guards who questioned them about what had happened. They got a stern warning about going into unknown waters and were eventually let go. The story appeared in the Cork Examiner and the boys felt like heroes but the parents were not too happy. However the boys thought long and hard, before they went boating again.

Boys being boys, we were always looking out for adventure and when daddy bought me a Daisy Air gun I was thrilled. This gun was not powerful, but gave me the desire to get a stronger one later. I did in fact buy two more as I grew up and during the Emergency I shot many pigeons, both wild and urban types, which helped supplement the dinner table when food was scarce. I used to cycle to Cork with the pigeons on the handlebars and sell them for six pence each. Joe Murphy later bought the gun from me and he asked me to mind it for him, as his mother would not allow him to take into the home with him. I did so and Joe soon forgot I had the gun, which I then resold him about a year after! Later, I progressed to .22 rifles. Just after the war, Italian .22 bullets became available. The Italian bullets were a washout. Some just fell out the gun barrel when fired and there were a lot of duds. Remington bullets were available before the war and they were excellent. Daddy stocked up some of these bullets, which again helped fill the table with rabbits, and other wild fowl. I used to cycle to

Midleton where there was a hardware shop, and I used to buy the bullets there. I was never asked for a licence, or even asked what my age was. Daddy and I used to go to the country and I got lots of shooting experience. Once when daddy and I were outside Midleton searching for rabbits we decided to get some target practice in. He spotted a white tin, close to a bush, about one hundred and fifty yards from the ditch in a field. Daddy rested on the ditch and took aim. He squeezed the trigger and the can went spinning away from the bush. Just then, there was a loud yell and a farmer got up from behind the bush, pulling up his trousers as he ran like hell. So did we! The poor man was having a quiet visit to the loo! We got into our Baby Ford and did not return to that area for a long time.

 I was now in my teens and feeling confident after gaining experience in working and earning some money to contribute to the family and also for myself. I had proved that I could stand up for myself and should now consider my future. Much as I wanted to, I felt I could not just hang around and do nothing useful. There was no work available so I started to consider the possibility of emigrating. Fate was about to take a big part in my future.

Chapter eleven

It was just as the Emergency was coming to an end that I decided that I wanted to go to sea. My brother Anthony had joined the Royal Navy and my sister Eileen was in the ATS (Auxiliary Territorial Service) in the British army, and she was stationed in Lisburn Northern Ireland so I had heard lots of exciting stories about what life in the forces was like. When Eileen was home on leave she smoked in front of daddy and we all thought this was so brave. On one occasion, she did not have a match to light her cigarette so she asked me to get a light for her and in return she gave me a cigarette. I went up to daddy and asked him to light the cigarette for Eileen.

"You light it" he said.

"But I don't smoke." I lied.

"Put that cigarette in your mouth and light it, as you have been doing for some time." he retorted. I don't know how long he had known of my habit but after that he gave me the occasional cigarette and I used to give him one, when I had them.

I began to seriously think of my life at this point. I'd spent many happy years growing in this beautiful seaside town, watching transatlantic liners visiting the port, studying the tall ships, which called into Cobh and generally witnessing so much of marine and naval life. I went to see these wonderful ships and boats at close range and listened to many stories, which the local salts and other seamen related about their experiences in foreign countries. The British Navy utilized the harbour in Cobh to its full extent, various types of warships had been anchored in the harbour and tied up at Haulbowline during peacetime days and so I had a fairly good idea of what I could expect.

It was strange that both my brother Anthony and I decided to go to sea and yet none of our known ancestors or relatives had ever done so before. After a short time considering my few options, I decided to give it a go and become a Radio Officer. It came as no surprise to my family when I decided to study and achieve my ambition.

Around November 1944 a friend of mine, who was a Radio Officer in the Merchant Navy, convinced me to take up the course. His name was Dudley Balburnie but we all knew him as Bal. He worked with the Blue Funnel Shipping Company and was on leave and back in Cobh for a couple of weeks. One day he came up Harbour Row, in full uniform, and everyone passing smiled with pride and pleasure at him. He looked very smart. He wore a navy overcoat with gold braid, a peak cap with the Merchant Navy badge, kid gloves, and full uniform underneath. I was envious! He told me he had qualified from The Radio Telegraph Institute in Tivoli, Cork and that the course and examinations were not too difficult. I was convinced that this was for me but I had to wait a few months for the course to start.

During this time, while Bal was at sea, another Cobh lad called Sean Carr, from the Mall, whom I also knew, was also at sea as a Special Radio Operator. Some lads had gone off to sea with the Marconi Company and were only qualified at sending and receiving Morse code. They had no electronic experience but due to the urgent need for ships to have some form of communication these Special Operators were allocated as a stopgap measure, until fully qualified Radio Officers became available. Also, it was mandatory to have them during the war. Later, these Operators had to come ashore, to sit the full examination.

I met Sean Carr at the Radio Telegraph Institute when I started in March 1945. He was studying for his full certificate. He used to relate stories about his trips at sea, with particular emphasis on The Rio Tinto, in South America. He loved talking about this and usually put on a South American Latin accent when telling these stories. Every day I cycled from home to Tivoli to attend classes. It did not matter if it was summer or winter, except that if it was exceptionally bad weather I got the train to Cork and a bus to the Radio Telegraph Institute. This was about fourteen miles each way and kept me more than fit in those days.

The Principal and owner of the Institute was P.J. O'Regan. He was very droll, and an excellent teacher who never seemed to lose his cool.

But before I set off to sea I was determined to find a girl and so it was that during one of my expeditions to the Railway station with the boys on a glorious sunny Sunday that I met true love. Dan, Mick and I were sitting on the wall awaiting the train when we spotted four girls walking and pushing bikes, as they came up from the Lower Road. Without hesitation we followed the girls and tried to chat them up. Eventually we broke down their token resistance and found out their names. Molly Kiely was the youngest, Eileen was her elder sister, Kathleen was their cousin, and they had a friend with them called Mary. Dan immediately moved in on Eileen, and Mick tried for Kathleen. From the very start Molly and I kept looking at each other and smiling. I remember Molly wore a light summer dress and had ankle socks on. Her hair was strawberry blonde and she had freckles across the bridge of her nose. She was slightly built and had a beautiful smile, and glistening blues eyes that twinkled. We struck it off immediately and she told me she was sixteen years old. I believed her at first, only to find out years later she was in fact only fourteen years old!

On the first day we met, I found out the reason the girls were walking with the bicycles was because Molly's bicycle wheel was punctured. She jokingly said I had done it to get talking to them. I asked her how I could have done it when they walked into my life with the bike already punctured but offered to fix it for her. Together, we all walked to the Baths Quay near my home and from there Molly and I went up to my house to get my puncture repair kit. We returned to the Baths, where I fixed the puncture and earned my brownie points. Whilst we were up at the house, daddy saw Molly and from that day forward always referred to her as 'Freckles.'

When the puncture was repaired we all went for a walk out the Water's Edge to Whitepoint. This was our favourite place whenever we had a courting session going on. I remember we formed into our own pairs, except for Mary who had nobody to hold her hand as there were only three of us boys. Molly and I found a little hideaway and held hands for a while, before I put my arm around her slim waist. She did not object so after a further while longer I put my cheek to her cheek. As we sat on the rocks looking out at sea we watched the cattle boat s.s Kenmare pass in front of us - very Romantic! We then got up and moved towards a little crevice in the side of the strand that was more private, where I kissed Molly for the first time. She had to stand on her tiptoes, ankle socks and all. At first she did not know how to respond, and later she told me I was the first boy that kissed her. Reluctantly we parted and the girls got on their bikes and headed back to the ferry at Rushbrooke to return home. I went back to my house well and truly smitten.

Molly was attending St. Angela's school in Cork City and I cycled each day to the institute for my Radio Officer training. This gave us plenty of time to meet after school. We went for walks to Fitzgerald Park, along the Mardyke and in Cork City we regularly went to the Savoy and Pavilion cinemas. This was lovely in the summer. Sometimes we had tiffs and Molly and I went our own ways. We were both strong willed and neither wanted to give in. Molly was a Leo and I am a Pisces.

On one occasion, when we were coming back from Fitzgerald Park it started to rain. Molly had only a flimsy summer frock on her and soon this was clinging to her body, her golden hair was dangling around her face. She looked a pathetic sight. We headed for shelter in St. Mary's Church, on the Quays, but the rain continued to come down. Molly asked me to take her on the bus home instead so she wouldn't have to walk. However, in my pocket I only had six pence which belonged to my father. Even at eighteen years old, I would not spend it without permission. I pretended to run for a bus a couple of times and made sure I did not catch it, even though I got wetter with each attempt. In the end Molly got fed up, and walked ahead of me, in the downpour, back to her house. She went inside and closed the door. All I saw as the door closed were the blue eyes, freezing me out. Not a word was spoken then or for about a week after. Then one day, whilst I was at the

Radio Telegraph Institute, the owner told me there was a phone call for me. I picked up the phone and I heard giggling from the other end. The conversation went something like this;
"Hello, who is calling?" I enquired.
Female voice; "Are you interested in Molly Kiely?"
"Who is this?" I asked, "And what business is it of yours"
Female voice; "I am a friend of hers and she asked me to ask you, if you are still interested in her."
"Please put Molly on the phone," I requested.
She replied that Molly was not there with her.
"I can hear her in the background" I said, and when there was no response I asked the girl not to ring me again at the Institute and I hung up. For ages Molly denied she was there but finally admitted that she was and was sorry for not answering when I asked her to do so. This tiff ended when Molly's twin brother Paddy came to Cobh for a day of fishing. He handed me a crumpled piece of paper, he was grinning and said Molly had asked him to give the note to me. Still grinning, and with a knowing look in his eyes, Paddy asked me if there was any reply. "I don't know what was in the note," he said, but I knew damn well he had read it many times between Cork and Cobh. Paddy was the salt of the earth. I was very fond of him and he carried many more messages between Molly and me over the years. Molly and I patched up our differences, but had more tiffs from time to time. We saw each other on occasions, separated and led our own lives, meeting new friends, but always having time for each other and coming back together.

After nine months study I nervously sat the Radio Officer examination. At the Morse test, a friend and fellow student of mine, named Maurice Cronin, was sitting next to me. The test started and half way through it, my pencil flew upwards out of my fingers, did a summersault, and landed right between my fingers again. Somehow I kept taking and writing the Morse and eventually finished and handed up my paper. Maurice told me that when he saw the pencil go up in the air he froze, and could not continue taking Morse and lost his place in it entirely. He failed the examination. Like all the other students I had to do the practical, written and oral tests and then wait, and pray, that I would pass. I felt awful about Maurice and I don't know whether he ever went on and qualified.

A month later, I was down in the Deepwater Quay with Dan Hunt and we were looking around out the Harbour. I had no thoughts of the examination when Willie Casserly, and his brother Paddy, who were both Radio Telegraph students came up to me and said; "Congratulations Lynch, you've passed the examination!"

I was flabbergasted, my stomach had butterflies flying inside it and I kept asking if they were serious. I had forgotten that the results were due out. I didn't believe them and got the next train to Cork where P.J. O'Regan confirmed that I had passed.

Eventually the big day arrived for me to be awarded my certificate. I was now guaranteed a position as a Radio Officer with The Marconi Marine Communication Company in Chelmsford, England. My name, along with the names of the other successful students, was forwarded to this company and we now had to sit and wait to be called up.

I asked P.J. "Why our certificates black and all previous certificates are were either maroon or blue?"

He looked me straight in the eye and said, without a smile or flicker of the steely blue eyes; "Mr. Lynch, I presume that the certificate is in mourning because of the low standard of education this year." I could only laugh at his droll reply.

Now, I was a fully qualified Radio Officer, licensed to go to sea in the Merchant Navy or to be attached to Coastal Radio Stations when I got the relevant sea going experience. I never intended to try for Coastal Stations.

Whilst I awaited a call from Marconi, I carried on as normal in Cobh. Dan and I were still enjoying each other's company and we carried on as any teenagers did in those days.

As the Emergency was coming to an end lots of things were slowly changing in Cobh. Liners began to return. Some rations were being lifted and we were all getting much nicer food. The first slice of white bread I tasted and loved after the Emergency I got from an American liner in the Harbour. It was like nothing I ever tasted before. It was pure white and light as a feather. The taste was out of this world. It melted in the mouth. I could have eaten a loaf of it. We got this by rowing out in a leaking punt to the liner at the mouth of the Harbour. We had to continuously bail out the water, which was leaking into the old punt. Anyway, some of the crew threw this delicious bread to us, along with some chocolates. Some liners were useless for giving anything to us and sometimes turned the hose on us, at the request of the local customs officer. We soon got to know which liners were most generous, and which of the customs officers to avoid.

Once, we went out to a liner and were looking up and waving at the passengers when we spotted a couple of lovely girls standing at the rail, near three nuns. We started talking amongst ourselves in Irish, about how lovely the girls were and that we would not mind having a date with them, or something like that. The next thing we knew was that the nuns started to give out yards back to us in Irish for our disrespectful conversations, telling us we should be ashamed of ourselves. Nobody ever rowed away as fast as we did. If it had been the Regatta race, we would have won hands down. We travelled quite a distance before we thought of baling the water which was building up in the punt. Our trouser' legs were soaking by the time we got ashore.

On another trip, we were a bit late coming in from the liner and there was a strong breeze blowing. It was difficult to row against the ebbing tide so we decided to try and 'sail' in. I had daddy's raincoat with me so using that as a mock sail we hoisted one oar to use as a mast, and the second was used as a rudder. The coat ballooned out with the wind and we took off at a rate of knots, cutting straight across the channel. We arrived safely and disembarked, thrilled with our ingenuity. It was very thrilling and exciting, and we felt invincible. Daddy's coat survived the ordeal too.

An episode in a rowing punt that same year did not go as smoothly. A local lad, Gerard Bransfield, and I took my sister Eileen and her friend Peggy O'Leary out in rough waters in a small leaky punt. Actually, nearly all the punts leaked so we never left shore without a few empty paint tins to bail the water out! With the tide full in we headed from the Bench, up past the Baths Quay. Gradually, the punt began to take in water and the two girls had to sit on bailing tins on the back seat. People on the quaysides were screaming at us to come in, but we carried on until we got as far as the Naval Pier. This was about a half mile from where we started and we had to return the same distance.

We turned around and nearly capsized as we headed back. People kept calling on us to come into the camber, but defiantly we continued towards the Bench. One soldier had stripped off part of his uniform and was ready to dive into the sea and rescue us. He kept running along the Bath's wall in line with us ready to jump in if he was needed. The Gardai now arrived on the Quay and watched as we slowly edged towards the Bench. All the time the girls were furiously bailing the water out of the punt and soaking from head to foot. Gerard and I were soaked, cold and very tired by this time. At last, in the darkness we reached the Bench, where a crowd of people were waiting and some had boats crewed and were ready to launch to rescue us. We were called all kinds of names for being so stupid. However, there was relief amongst them that we had made it safely back. None of the four of us was fazed. Even though none of us could swim we KNEW we were going to make it!

As luck would have it, daddy heard about our latest escapade and had come down to meet us. Strangely, despite the fact that I could not swim I have never been afraid of the sea, no matter what conditions prevailed, and I never felt that my final life struggle would be with the sea. Daddy told Eileen and me to get home, and that he would deal with us later. We were grounded and had to scrub and clean the kitchen floor, plus a number of other unbecoming jobs, for our punishment.

One of my punishments was that I got the dangerous job of whitewashing the walls at the back of the house. These walls reached the height of our three-story house and covered a very large surface area, stretching back the entire depth of the building. One of these walls belonged to Barry's next door, but this did not matter to daddy. He just told me to whitewash the lot. At this time we only had two rickety wooden ladders that had woodworm in lots of places. Daddy lashed the two ladders together with ropes but they only overlapped by about four rungs. It was a nightmare the top of the ladder because I had to hold a bucket of whitewash with one hand wrapped around the ladder, and then dip the brush into the bucket with the other hand. Then, I had to splash the wall as best I could. I had no visor, or protective glasses, and I got a lot of splashes into my eyes that burned like hell.

There was no way I could safely clean my eyes up there in such a precarious position. While I was up at the top of the ladder and there was nobody holding the bottom of the ladder I felt the ladder begin to slide to the right. I immediately froze and grabbed the ladder in a bear hug. It stopped sliding and I was petrified as I looked down on the concrete below. My voice did not function with fear and I couldn't utter a word. My body was rigid. I hoped somebody would come to my assistance but to no avail. After an unknown length of time I started to get my wits together and slowly, ever so slowly, I put one foot down to the next rung, whilst still clinging grimly to the ladder. It remained solid. Another rung and then another. Now I was more relaxed and descended eventually and safely to the ground, perspiration dripping down my back. After some time I went back up the ladder but only halfway up this time until I got more confidence back. The ladder was rickety, unsteady and bowed with my weight, and did not instil any confidence in me. How I never came off and fell to the concrete below I'll never know but I eventually finished that job. I was about fifteen years old.

   Years later, when I came home on leave from one of my ships, mammy said daddy had fallen off one section of the ladder but he was all right. I took one look at the ladders, saw the poor condition of them and took the axe to them. I broke them into smithereens, just as daddy had done with the barrel that Anthony had nearly drowned in. Daddy was not too pleased but he realised that I was right.

At the end of the Emergency the Royal Navy minesweepers arrived to clear the southwest coast of mines laid by the British during the war. These Royal Navy ships used to dock at the Deepwater Quay and were open to the public for visits on Sundays. On numerous visits to these ships I could not believe how slack security was. Visitors were allowed to go anywhere onboard without escort, and in most cases sailors were in their cabins with girls so they had other interests. I saw twelve .303 rifles and bayonets on an unsecured rack on deck in a corridor. There wasn't a sailor in view. After some time I took a bayonet as a souvenir, put it under my coat, and left the ship. Feeling guilty, I decided to return it to the ship, which I did as easily as I had taken it in the first place. I could just as easily have taken a rifle.

On another occasion, I was lucky not to have been crushed between the minesweeper boat and the Quay. Dan Hunt and I came on deck and when we were leaving we went to the bow of the ship, which was level with quayside. I stood on a bollard on the deck and jumped across the wire stays but caught my foot on the stay. My knee crashed on to the edge of the Quay. I felt myself slipping between the ship and the quay wall as I dug my fingernails into the concrete edge. Gradually, I pulled myself up on to the quay and rested with the sheer relief of having saved myself, only to watch my knee balloon up to a huge size. It was sheer bravado on my part trying this stupid game. I shiver now when I think of this. The quay and the minesweeper were only about two feet apart and with the swell in the sea, this reduced to about fifteen inches. If I had not held on to the quay, I would have been mashed between ship and quay. It would have been more sensible and easy, to walk off by the official gangway. If I had been mangled, there would have been plenty of eating for the crabs I used to catch.

The minesweepers stayed for quite a while and I took photos of them in the harbour. I gave the pictures to Charlie Nash, a local retired pilot who told me he later gave them to the Maritime Museum in Cobh. I remember the names of some of the minesweepers from then; Flagship HMS Ready, HMS

Crocodile, HMS Rattlesnake, and HMS Cheerful, all very innocent names for ships that were built to retrieve mines. Ironically, the crew of this latter ship, the HMS Cheerful, caused more fights in town than all the others put together. There were confrontations between townspeople, the Irish army and the British sailors. The civic guards and military patrols were active all the time trying to maintain the peace. The girls had a whale of a time though!

By January 1946 I was still awaiting a call from Marconi for my first placement. I got correspondence to say that the special Radio Operators were being replaced by qualified Radio Officers and that I would be called when a vacancy occurred. I waited, and waited, getting frustrated as time went by. I wanted to get away to sea, but I still had to wait a long time before my dream was to become a reality.

In the intervening time I did have an offer of employment, to work as an agent for the Irish Assurance Company for two pounds and ten shillings a week. I felt strongly that this was not for me so I turned it down.

In the interim, Dan and I joined the Local Defence Force. At this time daddy was in the Local Security Force where he was in charge of local transport and he made it my responsibility to record the names of everybody who owned any type of car, or mechanical means of transport. I knew every vehicle in Cobh and who the owners were.

In the Local Defence Force I got to experiment further with armoury. We were given .303 Lee-Enfield rifles which I kept at home. Neither bullets nor bayonets were issued. We met each week in the Hall in East Beach and had drill and instructions in handling, cleaning, and stripping the rifle. A serious incident occurred one evening in the hall when we were carrying out target practice, using blank ammunition. Six of us took up the prone position, facing six fellows, each holding a target disc which had a small hole in the centre of it. The lads held these discs to one eye and faced the fellow with the rifle. Behind them were more lads sitting on the bench against the wall. The idea was that the fellow holding the disc peered through the hole, to see the alignment of the rifle barrel facing him. We were separated by about ten feet. The instructor issued us with a clip of five blank .303 shells, and gave the instruction; "Five rounds, in your own time, fire."

I aimed and fired but there was only the dull click of the rifle bolt so I got ready to fire my second blank when there was a loud bang from the rifle second on my left and the sound of a bullet ricochet causing immediate pandemonium. My eyes automatically scanned left to the men holding the discs. No sign of blood! The Instructor shouted; "Stop firing and lay down your rifles!"

God only knows how the fellow holding the disc escaped. The live round went over his head, and took the gaiter off the left leg of the member sitting behind. Nobody noticed that live ammunition was mixed with the blanks. Talk about Russian roulette! Normally, the blank bullets were identified by colour code but were difficult to see when fitted in a clip. Fortunately, nobody got hurt directly or from the ricochet after the bullet hit the wall. A check of all the remaining bullets did not reveal any more live bullets and the only five were those issued to the unfortunate man who had fired the bullet. Both he and the man holding the disc were petrified. The poor guy seated was visibly shaken and it took some time to quieten him down. His gaiter was beyond repair. I don't remember any more in-house target training after this. The shooting of large calibre rifles normally only took place in the firing range so the sound of a .303 bullet being discharged travelled a long way.

Across the calm water, in Spike Island, the shot was heard and they were on the phone to the Garda barracks to find out what had happened. I never found out the result of the investigation. We were all shook up. I personally felt relieved because if I had received those live bullets there would have been a dead man in front of me. I was the top shot in "E" Coy 47th Battalion at that time and represented the Company at the inter-company shoot-off in Kilworth Army shooting range. I would not have missed from ten feet. The good Lord looked over all of us that night. Otherwise, it could have meant a nightmare life ever after.

We had some great times with the L.D.F. Every year we went to Fort Camden for annual training. We stayed in bivouacs - temporary tents - overlooking the entrance to Cork Harbour, close to Crosshaven. Dan Hunt and I shared the bivouac together and got on very well, we were inseparable. One night, it was lashing out of the heavens, when we heard terrible swearing as shadowy figures ran past our tent. We could see the shadow of a mallet being waved about. It was a couple of West Cork lads on the rampage, trying to catch some fellows who had loosened the stays of their tent. With the heavy rain the tent collapsed on top of them while they were asleep. The poor West Cork guys always seemed to get a rough time as the other lads classed them as 'Culchies', just because they were from the West of Cork!

Mealtime in Fort Camden was an unforgettable experience, for all the wrong reasons. We all trooped in single file past a soldier, who dished out soup from a metal bowl of what looked like sludge. A large lump of dry bread was thrown on the tray. There was also a cup of tea provided that looked like Guinness. We were so hungry we had to eat what we were given but it was far from what our mothers had fed us.

Getting washed in the mornings was another problem. As we were billeted in tents, we had to make use of makeshift outside cold water facilities. It was not pleasant, lining up for a chance to splash my face and shave. The outside communal loo was another eye-opener. It was nothing more than a square of grass surrounded by semi transparent hessian or canvas. Inside were a line of three open wooden loo seats over holes in the ground. There was no sanitation or loo paper.

The army never learned from the old days and lived by the old rhyme;

> '*In days of old,*
> *When Knights were bold,*
> *Before paper was invented,*
> *They wiped their ass with a blade of grass,*
> *And went away, quite contented.*'

Away from the regimented army living it was great to get time off to enjoy ourselves so we never missed an opportunity to go down to Crosshaven. This was a great place to go in the evenings and weekends when we were off duty. It was a very busy seaside resort with funfairs and plenty of talent to chase. Usually, we wore our mufti – the army term for civilian clothes - but sometimes the uniforms sufficed. To keep unruly members on the straight and narrow some fellows were appointed as Military Police, similar to Army Redcaps, to patrol the area and watch for unruly happenings. These men were identified by the white armband, which they wore. All who intended to stay out after midnight had to get a late night pass. Some fellows did not always do this and were arrested if caught out. One morning Dan and I were returning to barracks at one a.m. with late night passes in our pockets, when we heard noises from the side of the road. We realised that some LDF lads were in the bushes with some girls so we decided to play a trick on them. We put white hankies on our arms and shouted at the fellows to come out and show their late night passes. There was a moment of silence and then two guys came out and ran like hell away from us. We did not bother with them but instead approached the girls and pretended to question them about the fellows. It was not long before we had secured dates with the girls - nothing ventured, nothing gained!

There was another time when Dan and I were wearing our uniforms and were returning to camp and we met two other guys from Cobh who were going into Crosshaven on a late night pass. I could not believe my eyes and got mad as hell when I saw one of them was wearing my civilian suit. He said he only borrowed it as he preferred it better than his own suit. I did not believe that somebody could be so brazen and stupid. I chased him back to barracks and made him take it off, with a warning not to ever again take any of my property.

On still another occasion Dan and I met two girls and we walked them back to the house where they said they were staying. All was going fine until one asked us if we liked apples, as they knew of an orchard close by where we could steal some apples. We showed our bravado and said we'd have a go. In we went, while the girls kept watch and when we returned from our escapade with some apples one of the girls told us we had stolen from her parent's house. The two girls howled with laughter at our discomfort.

We had a regular army sergeant, named O' Shea, who took us for training most days in camp. This training usually took place in a big field nearby used by cattle for grazing. We had to run, fall into prone position, and do various other tactical exercises. Our biggest test was when we were running and a sergeant barked a command; Down! We had to drop where we were at the command. The field had plenty of cow pats so we all tried to avoid going down into one of these smelly piles of cow mess. Not everybody was successful and ended up with dung all over their faces, arms or uniforms. Some sergeants showed no mercy and selected spots where the manure was abundant, just for fun. Thankfully Sergeant O'Shea was not one of these tough soldiers.

We used to have great fun too playing tricks on new arrivals from the country. Sometimes, when we slept inside barracks, we made up French beds whenever a fellow was not around. This entails folding the sheets so that when a guy got into bed, he could not stretch his legs out because the sheets were folded back on each other. I heard lots of swearing when some poor fellow returned from a night out and just wanted to get into bed, only to find he had to make up his bed all over again.

We all had little moments when we would get into trouble or get up to mischief but it was always in good humour. Dick Leahy, from Spike Island, was a member of our battalion and he was out one night after midnight without a pass. He had been courting with a girl and started back over the fields to the camp in Camden. As he approached the camp, he climbed over a ditch and fell on top of a fellow and a girl entwined on the grass. He got up, brushed himself off, apologised and was starting back again when a stern male voice shouted, "Soldier, stop and explain what you are doing out at this time of night."

Dick looked and there was a regular Irish Army Captain, adjusting his uniform. Without hesitation Dick saluted and said; "Same as yourself sir," and then ran like hell.

Whilst I was in the Local Defence Force, enemy aircraft attacked the Irish coaster ship, Kerlogue, on her way back from Spain. When she came into the harbour, we were alerted to standby at the Deepwater Quay where she berthed and the Captain was taken away to hospital having suffered fatal wounds.

Later, during the war, the same ship arrived in Cobh, with German naval survivors picked up off the Irish coast. We were again alerted to standby and support. Regular army units arrived, and took the sailors away to the Curragh camp for internment. The Germans were all smiling, and I had the feeling they were delighted to be away from the war. One very small sailor had difficulty climbing up the back of the big army lorry and a hefty Irish army sergeant picked him up by the scruff of the neck and backside, and virtually lifted him into the back of the lorry.

The Emergency was now over and I'd had enough of waiting for things to happen. Marconi was tedious. I had come through the Emergency and had learned a lot about life and myself. I now felt confident that I could face emigration and when an unexpected opportunity presented itself to me to fulfil my dream I knew I couldn't turn it down.

Chapter twelve

By May that year I was fed up waiting for Marconi to call me and offer me a position so I decided to immigrate to England and work there. As it happened, Dan Hunt came to me one day and said there was an advertisement in the paper for recruits to the London Metropolitan Police. He asked me if I was interested, as he was going to have a go. The terms and conditions were that the Metropolitan Police guaranteed to pay the fare over for people in Ireland and, if for any reason you were not accepted they would also pay your return fare home. I jumped at the opportunity, and that week Dan and I got our return tickets from Cork to Fishguard by the m.v. Innisfallen, and from there to London by train.

 I left Cobh, my home, with fond memories and a feeling of trepidation at moving towards the unknown, as I had never been away from Ireland before. I said my goodbyes to the family and got the train to Cork. As the Innisfallen came down river and passed our house in Harbour Row and I watched my parents, sisters and the locals out on the street waving handkerchiefs, my mind went back to when I was younger, and how I had laughed at the emigrants on the boats leaving the harbour for good. Now, I had first-hand knowledge of how they felt and it was not pleasant, for the first time in my life I was lonely.

 The crossing on the boat was not too bad even though we had to sit on chairs for the whole trip. We could not afford a cabin. Fortunately the weather wasn't too stormy and the trip was enjoyable. We then had to rush for a seat on the train and we were lucky as numerous people had to stand in corridors for the entire journey. When we arrived in London Dan and I had to find our way on the Underground train and eventually reported to the Police station at Beak Street, off Regent Street. We stayed there overnight and started the acceptance process the next morning. During the first evening, one of the police men; they were called Bobbies there, told us various stories about Hyde Park and the ladies of the night, and how accommodating they were to the Bobbies. He spoke about the nightlife in general in the city and we thought; "This is going to be exciting."

The following morning we had our medicals, where we were checked for height, chest expansion, general health and fitness etc. I was doing fine up to that point but when I came to the eyesight test; I failed on the smallest row of letters. The letter 'Y' was my downfall. I was asked to return for a retest the following day but declined, as I felt that I would not pass the test that time either. I was not too bothered about this result as I really wanted to go to sea and felt I would miss the opportunity to see the world if I signed up for the police. Anyway, I held on to the hope that Marconi would soon call me for interview. There was nothing more to do but say goodbye to Dan and contemplate my next move. The full return fare had been refunded to me in cash as promised. I intended to seek a job in London.

Luck was with me. As it happened, Frank Harris, a friend from my early days, lived with his mother in Wembley and she invited me to stay with them until I got a job. I was delighted, as I did not know my way around London and was happy to have the comfort and security of a familiar face. The clothes I wore when I arrived are still burned into my mind, mainly because of Mrs Harris, who said when she saw me, "You'd never guess you were Irish, brown shoes and blue suit!" I was wearing a striped blue suit, white shirt, studded detachable collar and tie. To blend in, I wore brown shoes and a cream/white belted trench coat. In those days Irishmen were recognised in England by their style of dressing. I soon noticed that most of my nationality did in fact wear similar type attire.

Joe Murphy, a fellow friend from Cobh, arrived in London around this time too looking for a job and we met up. Joe and I headed off to join London Transport Buses. At the interview we were both called into some fellow's office and asked some simple questions like; "If a person paid you two shillings for a four penny fare, how much change would you give back?" or "Two people get on the bus and want two tickets - one at four pence and one at three pence. How much change would you give back, if you were given a half crown?" Despite us getting the answers right it turned out that Joe was too young to work there and I would not join without him.

Our next stop was the London Fire Brigade. Joe got scared when he saw the high ladders, and pulled out. Even though I was accepted, I too pulled out as Mrs Harris felt I was worthy of bigger things. I often wonder if I had accepted what my destiny would have been.

Now, Joe went looking for an uncle of his who lived somewhere in Cricklewood. Joe did not have any address for him and soon discovered that Cricklewood is a big place. I accompanied him as we walked the streets looking for a window cleaner, which was his uncle's profession. We kept asking people we passed; "Do you know an Irish window cleaner, named Brady, who wears thick horn-rimmed glasses?" Eventually we struck lucky and Joe was reunited with his uncle. Joe later found work there but many years later he longed for home and returned to Ireland.

Mrs. Harris was constantly scanning the newspapers for a suitable job for me and one day saw an advertisement in the paper for Murphy Radio in Welwyn Garden City in Hertfordshire. I went for an interview and medical and got a job as an electronic tester and inspector. There, I worked on equipment for the German Army. The main electronics were walkie-talkies. Some of the work was boring, as I had simply to check miniature transformers for continuity, open circuits, breakdowns etc. I had progressed to testing these Walkie-talkies soon after I joined. The people were very nice there though and I got on well with everybody. There were no smart remarks about being Irish, as was common at that time in England. All the employees were English and generally called me Pat. During my time in Murphy Radio I lodged in a semi-detached council house, very close to the factory. The owner was a small man, who was retired and smoked endlessly. His wife was a lovely quiet woman who was always looking out for my welfare and they had a grown up daughter who came to visit sometimes.

I celebrated my twenty-first birthday here but there weren't many celebrations. I don't have any memories to suggest that my twenty-first birthday was different from any other birthday. I greatly missed my family, who would have given me a hug and kiss had they been around and made it a special day. Many more birthdays passed in a similar vein, quiet and lonely except for the letters I received from home. They were always welcome and kept me in touch with my loved ones.

It was whilst I was here that I first tried to give up cigarettes. I was off them for about five months until one day I went into the Post office, to lodge money in my savings book, when a craving came over me. Without hesitation, I bought a pack of ten Players Weights. My first few draws were sickening but it did not take long to be back in the full swing of smoking again. It was to be many years before I tried kicking the habit again.

At long last, in June 1949, I was requested to report to the Marconi Offices in East Ham, for interview. I handed in my notice at Murphy Radio, said my good-byes, boarded a Green Line bus and sat back to enjoy the journey and looked forward to moving to London.

My first impressions of this depot were not too favourable. The chief guy there was named Mr. Dyer. He was a smallish man and had a limp. His attitude was very sharp and bossy. Whilst Radio Officers waited for appointment to a new ship, he used to call each one on a Morse key, using the ship's call sign, or by the individual's name if anybody was not attached to a ship at that particular time. There was enough Morse whilst at sea and all objected to this while ashore so nobody answered his calls. The guy just wanted to show off his prowess on the Morse key. It also saved him the short walk from the front office to the waiting room to deliver a message personally. Eventually, he gave it up and came into the room to call us individually. I got very bored on my own in this room so it helped, when other guys arrived to suffer the waiting game with me.

When I first reported to the depot I was required to go for a number of tests, to confirm my suitability to go to sea as a Radio Officer. I was a bit worried as I had not practised Morse code for a long time.

A Corkman, Mr. Dwyer, came in and said he was going to give me a Morse test. He was a slim man, going thin on top, aged about forty something. He started to send messages at about fifteen words per minute and gradually he sped up to twenty-five w.p.m. I found it a little difficult to keep up and was concerned when he took my paper to check the results.

"You're a little rusty from being ashore for so long. Just relax and we'll have another go. That was only a warm up." He said. Boy was I glad! The second time round I had no problems and did fine and relief spread over my body.

Mr Dwyer then said to me, "When you go on leave to Cork, remember me, and bring me back a nice piece of bacon and a bottle of Cherry Brandy. There's a lovely shop, named Maddens, in Bridge Street and you'll get both there." I found out later that a number of Cork Radio Officers used to keep our friend supplied whilst on leave; a case of 'you scratch my back, and I'll scratch yours.'

Having passed the interview and test, I was shown where to go to be fitted with my Radio Officer's uniform. My uniform consisted of a doeskin trousers and jacket, two white shirts, shoulder epaulets, black ties, black socks and white stockings, one black peaked cap with two white coverings for the tropics and white shorts. The final items were a scarf and navy blue raglan raincoat.

Marconi paid for all these items and the money had to be repaid by me from my monthly salary of £20.00. Marconi deducted the amount owed for the uniform and the balance due was paid in a lump sum when I signed off the ship.

The salary rose to a maximum of £44.00 per month, after a number of years. The most I remember getting was £30.00 per month, just before I resigned from the company after being with them for 5 years. They also furnished free travel warrants for home leave. I'm sure they made a lot of money holding on to these salaries. Some trips lasted over a year.

My next stop was to Her Majesty's Marine offices in London that June, where I received a seaman's identity record book. This book - also known as a Discharge Book - would contain my record at sea whilst I served on ships which flew the British Red Ensign Flag. This flag, as distinct from the White Ensign, was affectionately known by seaman as the Red Duster. I was also given a British Seaman's identity card. I already held an Irish Seaman's identity card, which had been issued to me in Cobh the previous May.

Experienced Radio Officers did not spend too long in any depot awaiting a ship. This was mainly because the Radio Officer was the only Officer of the ship's crew who had to be onboard before the ship could sail. Even the Captain did not have this distinction as the Chief Officer could act in his place. No Radio Officer, no voyage.

Inexperienced and first time Radio Officers had to wait for an opening to occur in a passenger ship so that there was a senior Radio Officer available to train him. I was now in this category and waited patiently to be appointed to a ship. During my waiting period I learned a lot from experienced Radio Officers whilst in the waiting room.

Watch hours were kept from 0800 to 1000, 1200 to 1400, 1600 to 1800, and finally 2000 to 2200 hours. All Watch times had to be kept and recorded in GMT. Of course some larger ships and tugs kept a twenty-four hour watch. During each watch period strict silence had to be maintained during, and between, the fifteenth minute and eighteenth minute, and between the forty-fifth and the forty-eighth minute of each hour. All Radio Officers had to listen in carefully during these critical minutes on 500Kcs. They were required to listen for, and respond to, SOS distress messages which might be sent and that otherwise could be missed if heavy traffic was being sent on 500Kc's. 'Q' codes were international, and were used to help alleviate language problems. They consisted of three letters, each beginning with the letter 'Q.' A few examples are;

QRA meant: 'What is the name of your ship or station?'

QRB meant: 'Where are you from or where are you bound for?'

After QRZ it followed on to QTA, and so on....

The broken red line around the circumference of the Radio Room clock is divided into twelve segments, and each segment is separated by one second. Each segment is timed for four seconds; this is to enable the Radio Officer to send accurate dashes, to set off alarms on other ships when a distress message was broadcast. Any four correctly timed consecutive dashes triggered the signal on any auto alarms on other ships. This immediately advised them that a possible distress situation was in progress. In between watch periods Radio Officers set their Auto Alarm in case there were any distress signals being sent. During watch periods they would have to listen for their ship's call sign in case anybody was trying to communicate with their ship. Also, weather reports and traffic lists would have to be listened to and copied on a regular basis, from coastal radio stations sending traffic lists, at their scheduled times. Coastal stations called up ships by call sign, in alphabetical order, advising that they had messages for them. Sometimes they would call ships outside of traffic list broadcast times. When this happened, the Radio Officer would have to reply to the coast station and change frequencies, receive the message, repeat it back if there was uncertainty of any part of the contents, acknowledge receipt, then change back, and listen to the distress frequency (500Kc's /600 meters.) Next, the message would have to be delivered to the addressee; either crew or Captain. If a new message had to be sent it would have to be transmitted to the nearest coast station closest to message destination. Sometimes, flowers would have to be sent to people ashore via Interflora. Some ships had ship-shore radiotelephones, and crew or passengers could call home. Most Morse communication was carried out at about twenty to twenty-five words per minute. A word for the unit of code counting was five letters. On occasions speeds of thirty words per minute would be broadcast.

 The position of the ship (QTH) had to be sent on a regular basis to the nearest coast station along the route. This position, and all communications held, plus daily time checks taken on sixteen Kc's, from Rugby Radio, in the UK, had to be recorded in the Radio Log Book. Normally, at least one entry was recorded in the radio log every ten minutes. This consisted of call signs of ships, or coast stations, carrying out traffic. It was compulsory that all distress (SOS), Emergency (XXX), and Navigation warnings (TTT) were logged, with times intercepted and any unusual occurrences during transmission. All such messages as well as storm, gale and other weather reports had to be delivered to the Bridge.

Quite often CQ (calls to all stations; which mean 'Seek You' messages were sent from various sources such as ship, or coast stations, asking if anybody had anything for them, or if they required information.

All equipment failures had to be recorded. Occasionally, the equipment might act up, and corrective action taken to repair the problem, and hopefully have the spare parts to carry out repairs. Other times, atmospherics could be so bad that it became impossible to decipher any Morse signals, and worse, the Auto Alarm would go off at all unearthly times. Sometimes it was necessary to get up three times, during the night, only to find it was a false alarm and that atmospherics were responsible. Sounds like a prostrate problem! Constant bursts of atmospherics soon fooled the equipment. It was absolutely terrible off the coast of Brazil, and while going down to Argentina on a voyage. These false alarms had to be reported to the Captain, and entered it in the Radio log. On some occasions due to continuous and prolonged atmospherics it was impossible to operate the Auto Alarm for some hours, and it became necessary to maintain personal radio watch.

Around coasts throughout the world, there are radio beacons to facilitate ships to navigate, when other means of obtaining the ship's position are either difficult, or not possible- fog, stars not visible, no radar etc. It was the Radio Officer's job to take radio bearings from multiple beacons, and then pass the information to the Bridge. Usually, a triangular inter section of the bearings was obtained, but if three stations were not available then an inter section of two bearings sufficed. A Direction Finder, using the Bellini Tosi system was utilised during these times. Beacons had their own individual call signs, and their exact locations were shown on the ship's charts. The Direction finder loop aerial was rotated through three hundred and sixty degrees, until the minimum signal strength was heard. This occurred in two directions one hundred and eighty degrees apart. A straight aerial was now used, in conjunction with the loop aerial, and the weaker of the two signals was the correct direction of the ship, from the station beacon. The beacon had specific times and radio frequencies, for sending their signals, so if a transmission was missed it was necessary to wait for the next transmission. Sometimes coast stations were used for Direction finding, by requesting them to transmit a signal to allow the direction loop to do its job. Charges were normally accrued for this facility.

Another duty was to test the Lifeboat radio survival equipment, on a regular basis. This was a fairly cumbersome piece of equipment, and needed a number of hands to operate it. It had handles, which were turned like the pedals of a bicycle, to generate power to transmit. It was held securely as it was quite heavy, and water proof. It was such a vital piece of equipment, that if an emergency arose, it was necessary to ensure it worked efficiently despite the hassle of testing it. All such test results, were recorded in the Radio logbook. The equipment was painted bright yellow for easy identification in the water.

It was also necessary to ensure that the ship's radio batteries were fully and properly maintained. To this end regular top-up with distilled water, and charging of batteries was performed on each battery. All such maintenance checks were recorded fully, in the radio logbook.

In the case of ships registered any place in the British Empire, there was a facility to send messages to Radio stations situated in any of these countries, (Canada, Hong Kong, Australia, UK, etc) and the messages were relayed free and only incurred charges at the final radio station. It was also possible to ask another ship to relay a message for you if you were finding it difficult to contact a particular station.

Sea life beckoned, and I was given a travel warrant to go to Southampton, to join my first ship. As I boarded the train for my destination, my heart was pounding, and I wondered what challenges life was about to bring. I had waited so long for this moment, and now this next exciting phase of my life was about to commence. I lay back in the seat as the train slowly chugged out of the station and headed for Southampton…

Chapter thirteen

My first ship was the RMS Isle of Guernsey. The ship was built in Dumbarton, Scotland by Denny Bros. in 1930. She was registered in Southampton, and owned by British Rail. Her tonnage was 2152.33 and she could reach speeds of up to 20 knots due to her twin screws and oil burning steam driven engines. The scheduled trips were from Southampton to St. Peter Port in Guernsey and then onto St. Helier, Jersey and back to Southampton. Captain Trout and Captain Pearce shared the Captain's duties on the ship during different voyages, and there was a crew of 63 including Officers. The total maximum passengers carried on any one trip was 1190. In the Radio Room I had a Marconi 381 transmitter, and a Cr.300/2 receiver. There was also a Direction Finder, and the very old type Spark Transmitter. Finally, Radar completed the list of equipment. The call sign was GQYJ. I Signed on in Southampton on 8th July 1949, and signed off in Southampton on 5th September 1949.

I was told that this ship had taken part in the evacuation of Dunkirk during WW2 and as a converted Hospital ship in 1939, had carried 839 wounded back to England. She was damaged and laid up for a while but later, in September 1942; she was converted to a troop carrying ship and carried six landing craft, and hundreds of troops, in the invasion of Europe in WW2. She was also the first ship to deliver mail to Jersey Island after the liberation of the Channel Islands in 1945. After my time on her I later learned that she was unfortunately scrapped in 1961.

So this was the ship with lots of war history I was to join on my first sea voyage to the Channel Islands.

On arrival at the ship a Steward took my suitcase to my cabin, which was located on the lower deck, close to other officers' cabins. Here it was; my home for an unknown length of time. The cabin was reasonable, as I found out from comparison with other ships which I later joined, but it didn't have a toilet with shower facilities. After settling in I returned to the Radio Room where the Chief Radio Officer, Pete Rosney, showed me the equipment and gave me a run down on the various procedures to be followed. Fortunately, most of the equipment was similar to the equipment I had trained on in the Radio Telegraph Institute, Cork so I found it fine and was comfortable using it.

I walked around the deck to get a feel for the ship and was surprised to find I was very relaxed, even though it was only about five hours before I was due to start sending my first messages to Niton Radio, as the ship left Southampton, bound for the Channel Islands. I soon got used to the routine of life on board and enjoyed meeting passengers, who came in for a chat and asked me to send their telegrams for them. During my entire time I only got sea-sick once and this was due to having eaten a salad and bending over quickly soon after. I never again got sea-sick on any of my ships.

Pete and I got on very well and each weekend, when he visited his friend in London, I used to occasionally help him get a plug or two of tobacco through customs for this friend. I remember the first occasion I was asked to 'smuggle' the tobacco for him. He went on ahead and boarded the London bound train and I put the two plugs of tobacco under my cap. After about five minutes I walked casually through the customs shed on the quayside. I was exiting the shed towards the train when a lovely young female customs officer came up to me and asked if I was going on the train. I said "No, I am just saying good bye to my chief." She smiled, looked at my uniform, and asked me if I had any contraband, or anything to declare. Feeling jittery, and beginning to sweat, as she looked me in the eye, I said "No."

Again, she smiled, and said, "Fine, go ahead."

With relief I went to the train and found Pete waiting. He asked me for the plugs. I took off my cap and surprise, surprise, no tobacco. Pete was looking at me in a funny way and asked me if I was all right, as I acted flustered. I told him about the customs officer and he put his hand out and took the plug of tobacco from the breast pocket of my uniform. It was quite visible to him and must have been to the customs officer. I had forgotten to put them under my cap after I had put them in my pocket on the ship. Boy, did I feel stupid! As I walked back to the ship the same customs officer again came up to me and asked if I had met my friend.

I replied, "Yes, thank you."

She was smiling, and looking at my pocket, as she said "good luck," before moving on. I smiled to myself, feeling relieved, knowing that she had let me off the hook and realising I had just learnt a valuable lesson about being more careful in future.

I really enjoyed my trips off the boat as the uniform worked wonders. There were always plenty of young ladies who appreciated my 'help' with information etc. They were not shy coming up and asking the most stupid questions, just to get chatting. Who was I to question their motives?

As was customary, all Officers dined at the Captain's table each evening. On my third trip I was sitting down to a meal in the saloon when a pair of arms encircled my neck and a soft girlish voice asked if I had made up my mind about what I was having. There was a giggle from the Captain, and from the other officers, as I realised I had been set up, and that the lovely female voice that had softly whispered in my ear – was the voice of the renowned 'Bubbles' and his cohort 'Daisy', two gay stewards from the Queen Mary, who were temporarily roistering for our regular stewards. They were very nice guys but I felt like an idiot being the innocent Irish boy not long away from Holy Ireland. It was on this ship that I soon learned the hazards of using a cutthroat razor on a rolling or shuddering ship. Up to this point I had always used one since I first started shaving, and soon found how awkward it was to keep shaving this way when the ship encountered heavy weather. Many a piece of paper decorated my bleeding face. Safety razors were my preferred choice from then on, until I purchased my first electric razor.

There were plenty of hazards for passengers on the ships too. There were many occasions when I saw passengers hanging over the sides of the ship, throwing up everything they had eaten. Seasickness was common amongst the passengers. The decks were a mess, and dangerous to walk on when people did not reach the side of the ship in time.

Once, while I was 'off watch,' I was strolling on deck and I noticed two men sitting side by side on a seat. There was a coat on the armrest between them. One of the two asked me if I could get him a drink of water and I asked him why he could not go and get it himself. He smiled and pulled the coat up, exposing handcuffs binding him to the other man. He was a policeman bringing a criminal back from Jersey. Life on board was certainly never dull. Indeed, one of the most exciting parts of life on the ship was the girls. There were lots of temptations with the ladies onboard. The uniform seemed to attract them like bees to honey. One particular girl I met was a nurse who was going home to England. She told me she had to take tablets to stay awake while working, and was now taking sleeping tablets to get to sleep. On this trip I let her have my cabin, as she could not get a cabin on board. Since I was on watch for the whole trip, she had the cabin to herself until morning. I was not aware of any regulations forbidding loaning the cabin and since it was a gesture made out of compassion and not for any reward I was comfortable with my decision. The steward did not know the situation and opened the cabin door to tidy up. I used to keep the door locked while on watch so he used a broken door vent to get into the cabin. She was fast asleep and woke with a start, screaming when she saw him. He beat a hasty retreat and was annoyed with me because I had not alerted him of the situation.

Another girl from Bournemouth came to Southampton twice looking for me. She had developed a crush on me after I had met her on deck for a little while before the ship docked. She said she wanted me to marry her. She was a peculiar girl, good looking but very possessive. I made sure I was 'away' when she arrived. On her first visit she caught me off guard and scared the life out of me with her persistent advances. I think she had a serious sexual problem, and I was not the guy she needed. Oh. It was a tough life!

Whilst ashore in Southampton I went to see Southampton F.C. playing in the old First Division league. Other times I wandered around Southampton, looking at the liners and large passenger ships tied up at the docks. Destiny is strange. I could so easily have been sent for training on one of these large ocean liners instead of my present passenger ship destined to ply the 'Home Trade', which meant ships operated within approximately 1000 miles of the UK. I had no regrets though because I continued to enjoy this experience and looked forward to many more variations in the types of ships I would later sign on, whilst pursuing a seagoing career. From the ship I worked Radio Stations at Jersey (GUD) and Guernsey (GUC) on 425 KCB's during voyages. On the mainland, I communicated with Niton Radio (GNI) on 464/480 KCB's.

I was sorry to leave after I completed my two months training on board the Isle of Guernsey. It was my first time then packing my clothes and books which was to increase fivefold before I ended up going to sea. I was quite excited, because I now had sea experience, even though it was only on a coastal run. Deep sea was to follow, with all its good and bad points as I was soon to discover.

Chapter fourteen

So, once again I was left waiting. This time I had to standby in the Marconi office in Southampton until a ship became available for me. After a few days I was given a travel warrant and told to report to the Hull office, where I was assigned to the Tramp steamer, s.s. Winkleigh.

I signed on to the ship in Hull, England on the 17th October 1949. The Master of ship was Captain Jones. Captain T. D. Jones was a small Welshman with whom I got on very well. The ship was destined to go to Africa to load Cocoa beans and take this cargo to Holland. I remember the trip as it was my first trip deep sea and my first time on my own in the Radio Room. What a responsibility! The fate of the crew, and the ship, could depend on me being able to perform competently if disaster struck.

Once signed on I settled in, got to know my equipment, and made sure that I had the necessary spare parts and paperwork etc. The morning we set out on the voyage I was on the bridge of the ship with Captain Jones and the Pilot. The First Officer was also there, and so was the able bodied seaman (A.B) steering the ship. We proceeded out to the North Sea, via a long channel bordered by mud flats on both sides. From what I remember, the channel seemed to be quite narrow, but on reflection so was Harbour Row, which I thought was very wide, when I was a child growing up. There were lots of wading birds on the mud and they made quite a lot of different noises. Eventually, the pilot disembarked and I took up station in the radio room. I contacted Cullercoats radio (GCC) and informed them of the ship's name and that we were leaving (QTO) Hull and proceeding to (QRB) Lagos, West Africa.

The Winkleigh steamed down the east coast of Britain, through the North Sea, rounded North Foreland, and came into the English Channel. Again I notified the requisite Radio stations of our position and where we were bound. The weather was dull, misty and quite cold. The grey sky made the water look dark green and uninviting. As we came out of the channel, I again sent the same message to French stations and I asked each station if there were any messages for our ship. I took all weather reports being broadcast.

    When we approached the Bay of Biscay I got my first bad weather report. A strong storm was brewing and we were heading right smack into the middle of it. The Bay is bad at the best of times, but I had one of the worst experiences of my life to date when we entered into the storm. Forty-foot waves pounded our ship, which was only loaded with ballast, and we were riding high in the sea. It was very difficult to maintain my balance and composure in the Radio Room and to stop my chair from skidding. Transmitting was almost impossible, as my fingers on the Morse Key were unsteady and sometimes slipped off the key. There was no way to tie down the chair, so I had to make the best of it. In the saloon it was no better. Plates, cups, saucers, and cutlery flew in all directions as the ship lunged pitched and shuddered in the high seas. The Captain called for more water ballast to help stabilise the ship. I picked up and answered three SOS messages from other vessels whilst we were going through the Bay. There was nothing the Captain could do as our ship was light and would have been of no real assistance. There were larger ships and tugs attending the distress calls. Some very heavy waves pounded our ship, and one lifeboat and the bridge door on the starboard side were badly damaged. It took us almost two days to get through the Bay of Biscay and we were all relieved and thankful that we had not suffered any really serious damage or causalities onboard. I had survived my first taste of really bad weather but was safely out the other side of it.

    Gradually, the weather improved and got warmer as we approached the coast of Portugal. During a 'Silence period' I broke a cardinal rule at sea. Due to a time error on my radio room clock, I sent a CQ message out on the distress frequency asking if anybody had a message for me. This is quite normal practice but NOT between the fifteenth and eighteenth minute, or between the forty fifth and forty eighth minutes of each hour. These times are classed as 'Silence periods' and are strictly taboo for normal traffic duties and especially on this frequency, which is open for Distress calls only at these times. At the end of the 'Silence period' I immediately got a call from Lisbon Radio Coast (CUL,) station admonishing me for my violation and requesting me to adhere to the 'Silence period'. I felt embarrassed and stupid, and knew then that I would be called to answer for this error at a later time. Sure enough, when we got back to the London, I had to give an explanation and was cautioned not to let it happen again. It did not for the rest of my time at sea.

Anyway, our trip proceeded without further mishaps or problems, and we steamed down the African coast. We passed quite close to Morocco, Rio de Oro, French West Africa, Sierra Leone, Liberia, Ivory Coast, Gold Coast, and onto Nigeria. Here we approached the pilot pick-up point at the mouth of the River and waited for a Pilot to assist the Captain navigate the ship up river. Pilots are required in most ports for this purpose and are specially trained for local navigation. Eventually, a small canoe came alongside amid a lot of noise and shouting. I could not believe my eyes. The Pilot was stark naked! He shouted that he was a Pilot and wanted to come aboard. Eventually, the Captain was convinced but gave the man a raincoat to put on him whilst he was aboard. As the tide changed the Pilot gave instructions in Pidgin English and I thought the Captain would have a fit as he watched his ship being guided up the river. There were all sorts of obstructions on the river; bends, logs, canoes, and the pilot merrily steered the ship through the murky brown waters without killing, drowning or maiming anybody. All the way up the river, the people stood on the banks of the river. They were mainly naked, or semi naked, and did not seem embarrassed as we gawked over the side of the ship. A lot of the men appeared tall, and quite a number of them held spears at their sides. Women and children waved and we waved back. Some of the women had terrific figures. Now I was getting my first education in what life was about. It was like human geometry; all shapes, and sizes, and this went for men as well as women.

The Captain and the ship survived and we eventually got to our destination, Warri, which looked like a village in the middle of nowhere, one hundred and seventy miles southwest of Lagos. Where we anchored might have been called Warri but there was no town visible. Soon we anchored in the middle of the river. Small huts and structures like wigwams dotted the surrounding area. There was a lot of activity as children played, and adults moved around the banks of the river. Small canoes soon started to arrive with cocoa beans. The beans were in large bags and the little boats were low in the water as the men paddled like hell to deliver them to the ship. They were paid for any bags delivered in good and dry condition. The bags were hoisted into the holds of the ship by derricks, and were neatly stowed away by natives who were hired by the Agents ashore. Fortunately, for these natives, the weather was good and there was no loss of beans due to poor sea conditions or rain. Later, in other ports I witnessed many loads of beans being discarded, when the beans were damaged by wet and windy weather conditions. When this happened the poor men did not get paid for their hard work. A lot of them were muscle bound for most of the year, and they had only a limited working life. On top of that they only got a pittance for their labour, so they could ill afford to waste their precious time, rowing the beans to the ship, only for the weather to destroy their efforts.

Canoes used to come alongside the ship at this port and I remember a woman named Mary who sold or exchanged bunches of bananas for a couple of pence or for food. Mary was small, and her skin was tight and dry from age and sun damage. The few teeth that she had, peeped out from leathery lips as she smiled. She came every day while the ship was at anchor. It was a common sight on the ship to see large bunches of bananas hanging outside the cabins, and everybody took one as they pleased. The bananas were quite small; about four or five inches long, but they were very tasty. I enjoyed a few every day.

One day, I decided to take a trip in a canoe with a young African boy, who for a shilling, said I could use the canoe for a day and he would guide me through the streams of the river.

At first I was a bit concerned, because the Natives did not look too friendly as my canoe moved through various inlets and past their huts. To see these big men, partly naked, some with spears, others with bows and arrows, so close to me and in the jungle filled me with apprehension. I think I had seen too many Tarzan films! I kept thinking to myself, 'Did anybody see me leaving the ship in the canoe?' 'Will I end up in a stew pot?' 'Will I be missed?' I thought of my family and home. My skin was clammy and I said an act of contrition, just in case. The further away we went the more I began to worry and so I asked the boy to return to the ship. It was frightening for me, especially with the trees overhanging the river and all kinds of weird jungle sounds of monkey screams and other animals and birds making their own noises. There was not another white man nearby and my mind started to get the better of me. I thought, 'No more, I've had enough.' We got back to the ship and the boy said he would come again tomorrow for me, when he would show me some more places and get me a woman. Needless to say, I did not take him up on his offer.

At night, I really worried because I could hear drums beating and weird incantations from the shoreline. I felt uneasiness at the sounds, and remembered all the head hunting movies I had seen, whilst growing up in Cobh. I wished I was back there now, safe and sound. The local shipping agent told me that there was Voodoo being practiced on land. That idea terrified me and if anything was to make my mind up about venturing out alone this put the final nail in the idea. One of the nicer parts of night was watching the fireflies. I did go ashore in the daylight, accompanied by other crewmembers, where we met two Irish Priests who were administering out there. They seemed to enjoy it and were delighted to meet a fellow Irishman. They tried to reassure me that the natives were harmless but this was my first contact with this world and it was so different to what I was used to.

During our stay here in Warri, the natives used ropes to climb up the side of ship and board it. The Captain spent time photographing the 'big boys,' as he referred to them while they climbed up the ship's sides, stark naked. He said, "Sparks, they won't believe me when I get home, so I have to have evidence."

When we were ready to leave, our Pilot came aboard and we started on our trip back down the river. It was no easier this time for the Captain. He could be heard muttering out loud, worried for his ship. All's well that ends well, and eventually the Pilot disembarked and kept the raincoat, and we proceeded out to sea. The relief was palpable from the Captain, as he set course and gave the order; "full ahead" for Lagos. The Captain kept uttering various Welsh prayers whilst he recalled his journey up and down the river.

Again, on arrival at Lagos, we picked up the pilot, and here we tied up at the jetty. This time the Pilot was decently clothed. I have only vague memories of Lagos itself, except for a few events which occurred. The locals also worked very hard to load beans on board. There was one young lad who was counting the bags as they were delivered on board and he seemed quite resentful when I checked to find out how many bags were shipped aboard. The Captain had asked me to keep an eye on the count, so I casually asked this young fellow the score. He looked up and said, "I have my matriculation and I can count, why you English think we are not educated?" I replied; "I never said you were not educated, and anyway I'm Irish, not English." His attitude immediately changed and I did not have any problems getting the information I needed.

Later, I went down into the hold of the ship and watched the bags being stored. I was amongst quite a number of natives who were working under the supervision of a foreman. His duty was to ensure that things were being done correctly. There was one fellow who seemed very lazy, and I asked the foreman why he seemed to be sitting down a lot. The foreman called him and said something in his native language, and I got a glare from the fellow. The foreman said that the fellow was dangerous and for me to be careful with him. The Captain was looking down the hatch, overheard this, and told me to come up immediately, which I was glad to do. It was hot and humid down there, with sweaty smells which were not too pleasant and that, plus the danger element, helped me to get up on deck quite quickly. I did not have to go down in the hold but I wanted to experience it and see what it was like.

A lot of white people worked and had businesses in Lagos then and it was common to see them in boats being rowed by natives. There could be up to eight Africans semi-naked, sitting on the gunwales of the boat, and they paddled large canoe type boats, as their passengers of white men and women sat facing them. The natives used to hide their 'Jewels' by pulling them between their legs, as they sat sideways on the sides of the canoe. However, according to the shipping agent, the natives used to open their legs slightly to expose the 'Jewels,' whenever there were women on board. He said it was comical to see the ladies pretending to cover their eyes with their hands, whilst in fact their fingers were spread generously apart to allow a good view!

After a hard day's work the natives washed publicly from a tap in the square near the ship. Their dark skin had a white coating of some kind of powder, which came off quite easily. Close by there was a small little public mini market, and lots of children were playing near their mothers, who were busy trying to sell fruit and vegetables. They used old wooden boxes to display their goods. I was fascinated with one very large lady, who wore a skirt and what appeared to be a very large man's vest. Whenever she bent over to pick up something from the box, her right boob fell out due to its size and weight. I had never seen anything like this in my whole life, colour, size, or weight. Our time in Lagos had come to an end, but not my education.

We next headed for Accra, Winneba and Tacoradi, in Ghana. In all these ports we continued loading but I have few recollections of what events happened except for one incident in Winneba. We were tied up at the Quay but on the opposite side to the Quay the 'Bumboats' came alongside, and they usually wanted food or whatever we would give them. Bumboats are small rowing boats or canoes and were common in most ports in Africa where the natives traded with ships. Generally there was a mixture of men, women, boys and girls in these boats. They could be very persuasive, and crewmembers bought goods or exchanged cigarettes and sometimes threw coins or food to them. They were marvellous swimmers and divers and they had to go into the water every so often to retrieve badly aimed gifts from the crew. It did not take the crew long, on this occasion, to notice a beautiful young girl in a white vest, who dived over a couple of times, and when she came up her beautiful body was totally visible through the transparent vest. The lads then got a full loaf of bread and lowered it over the side just outside of her reach and made gestures for her to remove her vest before they would give her the loaf. No way would she do so and the boys were very disappointed. Eventually they gave her the loaf.

I did not go ashore in any of these ports as they did not seem to offer any attraction. However I did make two purchases in these places. The first item was a pair of wall mounted buffalo type horns and the second was a very light carved wooded paddle. It had the name Margaret carved on it but the name was misspelled. These two souvenirs later caused me some grief on the way home in London.

Next we arrived in Freetown, Sierra Leone. The weather was very hot and sometimes humid. The usual native activities occurred. Some of the crew bought monkeys, which they intended to bring home with them. The animals were wild and vicious and difficult to control. One escaped and spent part of the trip up the mast swinging off my radio aerial. Despite all attempts to catch him nobody succeeded until late into the trip. On one occasion our pet cat was on deck and did not see this monkey watching him from the mast. Slowly, the monkey descended and crept up behind the cat. Before you could blink the monkey had mounted the cat, and there was one hell of a meow from the cat, as claws and teeth came into action. The poor cat eventually escaped intact and disappeared indoors for safety not venturing out on deck for days. Eventually, it cautiously came out on deck with its eyes rolling, obeying the rules of the road; he looked left, looked right, and looked upwards, before summoning the courage to sit down, whilst maintaining his vigilance all the time. When he spotted the monkey his eyes never strayed from him and he disappeared the moment the monkey moved. This saga carried on until the monkey got hungry, came down to eat, and was eventually recaptured.

The Second Mate owned this monkey, which created havoc in his cabin. I was in his cabin one evening and I asked the mate what happened to his pillow, which had brown skid marks on it. He laughed and said the monkey used it like we use toilet paper. This was too much for me. The Mate did not seem to mind and thought the monkey was great fun. Nothing I said to him about the state of his cabin seemed to ruffle him and I did not go to his cabin again for the rest of the trip home.

I was glad to be heading towards home but first of all we had to go to the Canary Islands, for bunkers; which is fuel necessary for the ship's engines. The Chief Steward also had to collect some items for our upcoming Christmas dinner.

When we arrived at the Canary Islands the first thing we saw at the dockside was lots of locals trying to sell actual canaries. We were advised against buying any, as we were told the birds never again sang when they left the island. This was because the birds that sang, and which you thought you were getting, were switched by the locals when you bought them and you got a mute which did not have a note in its head. The lads were told by the sellers that the birds would be silent for a while until they got used to their new environment and they would then burst into song. Despite this warning a number of the crew bought birds and cages but forgot about birdseeds. They still hoped the birds would sing before we got home. During the trip two died, and nobody ever heard a whistle or warble from any of the birds. The sucker crewmembers did get another bird from the rest of the crew.

In order to prepare a Christmas feast to remind us of home the Chief Steward went ashore and tried to buy turkey, ham, and any nice goodies to celebrate Christmas on the way to Amsterdam, which was to be our next port of call when we left the Canary Islands. I don't know how hard he tried but the end result was all he could get were six scrawny chickens, malnourished, and not looking much bigger than canaries! He had to barter two bags of potatoes for these specimens but said that with plenty of feeding they would look like turkeys when we got to Christmas day. What a Hope! The birds were kept in the storeroom and were fed all kinds of food. Two died within three days.

As we passed through the Bay of Biscay the Captain advised the crew that they would have to put the monkeys and birds into quarantine on arrival in the UK. This had a profound effect on the owners and I'm afraid the poor animals were disposed of in the Bay of Biscay on the trip home. There were some sad scenes as the crew and animals parted company. The monkeys were fitted with some of the ship's stock of metal shackles and then dumped into the Bay.

The weather through the Bay this time was reasonably calm but very cold. As we proceeded through the English Channel it became bitterly cold and we got our share of rain and snow. Ice formed on all parts of the ship and it was dangerous on deck. We celebrated Christmas day at sea. We had three small chickens between the officers and crew which totalled approximately 30 men. From what I remember I got a piece of skin and that was it. The rest of the dinner was no different to what we had been served during the voyage. It did not compare favourably to the Isle of Guernsey, where we had silver service but then that was a passenger ship and the Winkleigh was a tramp steamer. Also, at this time, I had not tasted the food eaten on other ships. I did find out later that the food on the Winkleigh was average compared to most British tramp ships but far behind other ships, foreign and American owned, which I later served on.

Less than one week later we arrived safely in Amsterdam and discharged our cargo of Cocoa beans. We were in port for approximately three days, plus New Year's Day. Amsterdam was an eye-opener for me. I spent every day ashore. I had never seen anything like it and loved it immediately. There were beautiful souvenir and gift shops. In one large shop I bought my first gifts and presents to bring home - a statue of a Dutch sailor and some cushion covers, the usual first time buys at sea. I saw these souvenirs in lots more ports, and on bumboats throughout the world but this was my first opportunity to bring something home from abroad. I wandered around the canals and wondered at the number of bicycles being ridden around the city. I was gob smacked by the speed at which cars were driven, and it confused me for a while because I was not used to seeing traffic being driven on the right instead of the left side of the roads. I soon got used to it.

The Royal Palace in the centre of Amsterdam was a beautiful sight and I went there a few times, hoping to see a glimpse of the Royal Queen but I had no luck. Whilst there I had a couple of strolls (only!) down the Red Light area, and this was something else. The windows were decorated with all types of females, in and out of clothes. I was too afraid to take a chance on going into one of the shops as I had been hearing terrible stories about VD, Syphilis and other diseases on the ship. Some said you were not a sailor until you got VD nine times. Being cautious paid off, as a few crew members caught a dose and had to be treated. It was frightening but some fellows continued to chance their luck.

We left Amsterdam and steamed up the Thames, where we docked in Popular and I signed off, on the 5th January, 1950. It was here in Popular that I first got my British Seaman's identity book in June 1949 and here I was, back again. I reported to the East Ham depot and handed in all my logs and paperwork for the trip. After completing the necessary formalities I was eventually called by Mr Dyer. He paid the salary due to me, at £20.00 per month, less one shilling per month, which was given to some charity or other. I was also given a travel warrant to go home on leave for two weeks.

I packed my bags in record time. Thrilled, delighted, and fully loaded with cash I boarded the Underground at East Ham and got the first available train out of Paddington for Fishguard, to board the Innisfallen for Cork. I posed a severe hazard for people on the Underground and all the way home, as my African souvenirs were seriously dangerous instruments. I had a sharp pointed wooden paddle, about three feet long, tied to one suitcase and a pair of large decorated bull horns strapped to the other case. One suitcase in each hand was difficult but it was almost impossible going down an escalator full of people or on a crowded train! Many got prodded, and I got sworn at several times before I safely stored the bags when aboard the transport. I swore that never again would I be caught out like this.

    My parents, and Molly, were not aware I was on my way home as there were no phones in my home and communicating between England and Ireland was by letter or telegram. I usually wrote home from abroad but there was no time for this right now because I reasoned I'd be home just as quickly as a letter would arrive.

    It was strange sailing up the river Lee, past our house, watching all the people waving at the Innisfallen, knowing my family were not aware that I was on board. Of course Molly did not know I was onboard either and so was not waiting to see me in Cork, when the boat docked. There was no time to call in to Molly as I wanted to get home quickly. I was in full uniform and feeling proud as a peacock when I got off the train in Cobh. I was not expecting the embarrassment I was soon to cause to a friend and my enjoyment which the episode was going to give me.

    At that time, the Irish Naval service had a lot of young cadets whose uniform was the same colour as mine, and they too wore peaked caps. It was common to see army Redcaps (Military Police) on duty at the railway station, ensuring the best behaviour of Irish marines and army personnel, boarding or alighting from the Cork trains. The Redcaps always sprung to attention and saluted when Army or Naval Officers passed. As I approached the barrier I saw Mr. Mac Nicholas, an Army Redcap sergeant whom I knew very well, was on duty at the gate. He did not appear to recognise me in uniform and apparently only saw what he thought was a Naval Officer approaching. He immediately sprung to attention and saluted. I stopped, looked at him, smiled and said,

"Hello, Mr. Mac."

He gawked, and I can't repeat what he said when he recognised me, especially as I wore a British Merchant Navy cap badge. Enough said! We laughed later.

There's a typical Cork greeting when people meet you after you've been abroad. Without fail a friend will ask you the following "Are you home again, when are you going back?" It never fails to amuse me. What the hell! I think it's a grand salutation!

I left the station and walked home, dragging my heavy suitcases. People stared at my souvenirs as I passed and must have thought I was from another planet. It never dawned on me to get a sidecar or other transport, as I had never been used to doing so in the past. When I walked along Harbour Row, I saw Sparky, our black and white mongrel terrier, sitting near the door of my home. When I was about one hundred yards from the house, he jumped to his feet and ran towards me barking with joy, with his tail wagging. I had not said a word or made any signal. I dropped the suitcase and hugged him. There was great excitement when I went into the house. Mammy was overjoyed in her own way.

She made a pot of tea and got her woodbines out and sat while I told her about my trip. Dad came in from work later and was totally surprised to see me. He could not get over my uniform and was very, very proud of me. Anthony was away in the British Navy, and my sister Eileen was married in England. My other sisters came in one by one and we had great chats, and they all got some gift.

I went to Cork to meet Molly. She was in the Bons Hospital having had her appendix out. I had no idea of this until I called to her house and her twin brother Paddy told me about her. I bought a box of chocolates and headed for the hospital. When I got to there I asked for Molly's ward and bed number. The young nurse looked at me, gawked, and then ran down the corridor without even telling me where Molly was. I followed, and was met at the door by four nurses staring at my uniform and calling Molly. There she was, looking wonderful and fit. We hugged and had to pull the curtains to stop the glances being cast our way. Molly told me later that the nurses asked all kinds of questions about us and she was the center of attention until she was discharged. I called to see her again and we enjoyed our times together, and even went out some days when she was strong enough. We had no idea at this time how our futures would merge so closely. As time went by we began to share experiences we had in our individual lives and move closer to each other.

After about twelve days I got a telegram from Marconi, which requested me to report to East Ham. My leave had not yet expired and I was annoyed at being called back early. Being inexperienced, I reported as requested.

Chapter fifteen

After arriving in East Ham I went through the usual formalities. Here, I had to wait for four days before being assigned to my next ship, s.s Tower Hill. I can't remember where I stayed in East Ham whilst I kicked my heels. At least I did not have to travel out of London for my new ship as she was berthed close by in Tilbury Docks. Of course, I did not forget my Cork friend, and brought bacon, and Cherry Brandy to him. He was thrilled, and we had a good chat about Cork.

The s.s Tower Hill was registered in London and had a gross tonnage of 7258.17 tons so was one of the sturdiest ships around. Her captain was Captain L. Luishman, Master of the ship and he, along with the Second Mate P.J. Morley and the Third Mate Cyril L. Longhurst were an accomplished crew.

I signed on this ship in Tilbury, London on January 23rd, 1950 for what was set to be a six month voyage. We proceeded from Tilbury into the North Sea and set off to await orders. Everything was running very smoothly with the only discomfort being the cold weather. As we got into the English Channel the weather began to act up, and we had some nasty heavy seas and very heavy sea traffic. Ships seemed to be all around us going up and down the channel, as well as the ferries criss-crossing between the UK and France. Strict watch had to be maintained at all times, as we did not have radar on the ship. I was wondering what the Bay of Biscay would be like this time as we were heading for Gibraltar through the Mediterranean, and on to Port Said in Egypt for orders, where our final destination would be revealed. It was typical with tramp steamers that they are ordered to sea by their owners with no final destination having been directed. The ships would receive their orders whenever the company got business and that could be at any stage along the voyage.

I kept taking weather reports on a regular basis and the weather forecasts looked to be improving. When we did get into the Bay it was smooth like glass. We made good time and were soon going through the Straits of Gibraltar. The North African coast was soon visible I could see that Gibraltar was very close. It was a beautiful sight. The weather was glorious and I spent every free moment I had out on deck watching the coastlines and the ships passing by. I saw flying fish as they flew and glided across the sea, often landing on the deck of the ship. The ship's black cat had a feast when the fish landed at his paws. It was all strange, new and exciting to me and I enjoyed every second of each day. I had been communicating our position regularly with Alexandria radio (SUH) This radio station had a high squeaky tone. It seemed to go up, and down, as its signal was being transmitted. It was very distinctive and easily identified which made my job slightly easier. Eventually, the coast of Egypt came into view and we got our Pilot and docked in Port Said where we tied up to await final orders.

    Not having our orders yet was great as it meant we had time off to go ashore and do some shopping. There were many bumboats around the ship and all kinds of things available from these people. Many of the Arabs came up to me and asked if I wanted any 'Spanish Fly.' They showed me a small bottle, alleging it contained Spanish fly, what they explained was a potent aphrodisiac. To me this meant nothing at the time as I had never heard of an aphrodisiac. They also tried to sell me 'genuine jewels,' as they put it - allegedly just stolen that day from a rich person's house. These jewels always turned out to be glass imitations. Homosexual suggestions were also widely on offer but I don't know if any crew member accepted these. Suitcases were a big sales item. Beautiful 'genuine leather cases,' which turned out to be compressed cardboard and disintegrated in wet weather, quite ridiculously unsuited to a life at sea!

    I was certainly learning about the raw part of life, which I had never seen in Ireland. Cushion covers of very rich colours and materials, highly decorated with gold and other coloured threads sold like wildfire to the crew. Ashore, there was very good value in genuine leather suitcases, moccasins and various other items. Also on sale were shirts, and other clothes, as well as material for making suits. Perfumes were also to be bought easily, as well as all kinds of bric-a-brac. It was best to shop ashore as some of the local bumboat merchants were clearly conmen.

One the second day I went ashore to post a letter home. I was directed to a large enclosed square where Egyptian post office people were dispensing stamps, and other items, to queues of customers. The Post officials were behind grills, which had an opening about six inches clear from the worktop, and about one foot clearance at the top. When I approached to be served, I asked for stamps - which I had been advised to ask for by the Shipping agent. I had English and American coins with me and offered these. The fellow took them and made some calculations before giving me the stamps and some Egyptian money as change. What happened next is beyond my comprehension, even to this day.

I was lucky not to have been put in jail, or beaten up. I felt that the teller had somehow overcharged me and given me the wrong change. Why I thought this I have no idea, except that it was rumoured that these people could not be trusted. But I was not going to let him away with it. I argued with him and called him a cheat. He spoke good English and became furious with me for my accusations. All this time the people in the queue were watching in amazement. The teller eventually got so frustrated he picked up a handful of notes and threw them over the grill. They floated down around me but I did not touch the money. I was just considering my next move when I felt something hard in my back. It kept jabbing and as I slowly turned around I saw an Egyptian Policeman, in black uniform, complete with Red Fez and black tassel. He stood about six feet tall and had a black bushy moustache. Worse still was that he held a .303 Lee Enfield rifle, which was causing the discomfort in my back. There was not one sign of emotion on his face and his brown eyes stared through me. He nodded towards the gate and I thought, "I'll be locked away in an Egyptian jail to rot and nobody will know I'm here." Fortunately, when we got to the gate the Fez gave me one last hard jab with the gun, and pointed to the docks. Not a word was spoken by either of us. I tried to keep cool but thought, "He'll shoot, and say I tried to escape." Perspiration ran down my back as I walked back towards the ship, without stamps or change. I resisted the urge to look back, and prayed and prayed. At the first opportunity, when I got out of sight, I ran faster than I ever ran in my life and did not dare set foot off the ship again in Port Said. This terrifying incident should have taught me to be much more careful in foreign ports but unfortunately I was young and felt invincible and further scrapes and episodes reveal it took me a long time to learn my lesson.

Whilst on board there were numerous sexual approaches to crew members and to me by homosexuals. This happened again on later voyages and must be part of the culture in Egypt. On another occasion, whilst I was chatting to the third mate, Cyril Longhurst, I was overheard talking and one Egyptian jumped up, pointed at me, and shouted, "You deValera's man." I was dumbfounded and could only laugh. He wanted to shake my hand and seemed genuinely pleased to see, 'deValera's man.'

Another problem we encountered in Egypt was that some of the fellows used to come aboard to flog souvenirs but once aboard they tried to steal anything they could. They were experts at this and I soon learned to protect my belongings. Sometimes a hand would come in through portholes, to take anything visible on the bunk, or they used a long implement with a hook to reach other stuff visible to them. You could never really let your guard down.

Bargaining was a way of life here. I never paid top dollar for anything as I could usually get it much cheaper than the price asked. This bargaining experience came in very handy on later trips.

Eventually, three days later, we got our ship's orders to go to Russia for a cargo of grain but later that day the orders were changed again, due to the outbreak of the Korean War, as it was deemed unsafe to go to Russia. We continued to await further orders and in the meantime the Captain decided, as labour was cheap, to have the ship's rust chipped and then repainted in Port

Said. Gangs of Egyptians started the chipping, and were seated on planks lowered or raised along the sides of the ship. As they moved from bow to stern they had to be careful, because the toilet and galley outflows were located along the sides. On one occasion the Chief Engineer went and had a crap in his toilet and released it on top of some unfortunate fellow who was directly in line with the outflow. I happened to be watching at the time and felt sorry for the fellow but some of the crew laughed, and shouted "Well done Chief." These poor fellows were not allowed to use the ship's toilets and so a makeshift exterior toilet was erected for them at the stern of the ship. A few planks were extended out from the stern and this makeshift platform was given privacy by a canvas lining wrapped around it. The workers had to go out, squat, and do their business into the water below. The boats below gave this area a wide berth.

Eventually, after a further 4 days the agent came aboard and brought orders for us to proceed to Freemantle, Melbourne, and Geelong, in Australia. We were to pick up a cargo of grain. Stocked up with provisions, bunkers, and other necessities, we prepared to join a convoy through the Suez Canal. Searchlights were placed on the bows of all ships proceeding through the Canal for safety's sake. Each ship had to carry a licensed pilot. These were usually British but sometimes French or American pilots were also on duty.

As we set off with the convoy towards Suez we were aware that there was always another convoy leaving in the other direction from Suez for Port Said, and passing convoys would respect each other's space as they went by in the Bitter Lakes. Due to scheduled timing one convoy arrived in the Lakes first and would stop and allow the other convoy through before they proceeded onwards, as the canal was not capable of taking two ships abreast at any one time.

The convoys could consist of a mixture of any, or all, types of liners, cargo ships, tankers and sometimes warships. A ship could be placed anywhere in the convoy. Most Scandinavian ships were usually beautifully maintained, well painted, and clean, whilst some of the other ships from some countries looked rusty and weather beaten. There was a British Army barracks on the banks of the canal and the soldiers had plenty of spare time to watch the comings and goings of these convoys. The lads sat down on the banks and shouted abuse and foul language at any British ship which was in a poor condition. I was on one ship that was badly in need of a paint job and it was flying the Red Duster. She was sandwiched between two immaculately painted white ships as she moved in convoy. As we passed the British Soldiers they started their tirade of abuse.

"Paint that F***ing rusty heap of scrap. You're a F***ing disgrace to the British flag."

The Captain was livid, and shouted back "You scumbags" and other cordial greetings.

This happened to lots of British Flag ships. This time our ship had been repainted so we got a cheer of admiration from ashore!

Eventually we went past Ismailia and headed to Suez where we got rid of the searchlight and the pilot disembarked. We were now on our own and we headed through the Red sea bound for Freemantle, Australia. I carried out my normal radio duties and sent our departure and destination data to the local Coast Radio stations, and to the UK and Australian stations. Later, for the duration of the trip, I kept listening for messages, distress signals and weather reports.

However, there was to be a very rude awakening soon after we left. One morning, after the Mate on the midnight to four a.m. watch changed, the Mates exchanged pleasantries as one relieved the other from duty. All seemed well but it was still dark and we had no radar. The Mate looked at the compass and the course was correct. He looked at his charts and then came out on the wing of the bridge, into the open air just as dawn broke, to correlate the data with what he could see. As he squinted and stared into the horizon he could see something far in front of the ship. He couldn't make it out though. It looked like a dark cloud but was very low down. He felt uneasy so got his binoculars out for a closer look. Two seconds after adjusting the lenses he nearly soiled his pants with fright. What he thought was a cloud was actually land, and the ship was heading straight for it. He immediately stopped all engines and called the Captain to the bridge. They could not understand what was happening. There was not supposed to be land on their course, the charts had nothing on them to indicate same but unless their eyes were deceiving them there was a land straight ahead which they should be passing well to starboard. They immediately took bearings and found the ship had somehow gone off course during the last watch. They immediately reset the course, said a quick prayer started off without grounding or damaging the ship. It appeared that the man on the wheel had dozed off and did not notice the change of compass setting before eventually waking up and readjusting the course. The Mate on watch had not noticed the error either. It taught me that there is no point fully relying on data, instruments and charts unless you make sure that all the information is supported by what you can actually see in front of you.

We were now going across the Equator and the normal routine was for any first tripper on board to be treated to a rough time by the crew as we passed this important milestone. This could include stripping the person naked and dousing him with slops from the galley, hosing him down, or any other number of unbecoming treatments. Thankfully, nobody knew that this was my first trip south of the Equator line but still I sweated until we finally cleared this obstacle.

The rest of the trip was beautiful. We did not have any bad gales or storms that I can remember. I had my first sighting of an Albatross. It was huge and graceful as it glided through the air. The flying fish too were a sight to behold and the sunsets on the horizon across the wide ocean were breathtaking. I loved to watch the crimson ball of fire slowly sinking into the line between sea and sky. The ball would gradually take the form of a semicircle, and grow smaller as it dipped lower and lower. At the last moment, when it just started to disappear, all I could see was a greenish blue glow as the ball of fire reflected through the sea. There were colours ranging from scarlet, to gold, and blue and green. It was truly awe-inspiring.

Eventually, the time had come. We arrived at the Pilot pick up point in Australia. I had already transmitted our ETA (estimated time of arrival) by radio so the pilot was ready to board. I now sent my signal to all stations, letting them know that we were entering port and asking for any messages before I closed down. Signals were acknowledged and I was now free to do as I wanted until we left again and dropped the pilot off.

Freemantle was a nice port but we did not have much time there. I did however have time to visit Perth to have a look around but only after the ship was cleared of all inspections on arrival where all had to go through customs, immigration, and undergo a medical for VD. The medical entailed what was termed a 'short arm inspection' - in reality a visual sexual inspection - that strangely only applied to ordinary sailors, able-bodied sailors and crew members but not officers. I could never, in all my time at sea, understand why just the crew members should be subjected to this examination, as some of the officers also went with women and could easily have contacted a sexual disease.

We all got a clean bill of health but first immigration wanted to check any animals on board. The chief steward was responsible for the cat, which was a full-grown black Tom. The Immigration Officer told the steward that he would be held responsible for presenting the cat before departure. All was fine until we were ready to depart and calamity, there was no cat. He was ashore doing whatever Tom's do ashore. Panic ensued in the Steward's department. There was still no sign of Tom even after exhaustive searches so the decision was made to go ashore with the crew and endeavour to capture a big black cat. Any black cat would do. Unfortunately, for some reason, there appeared to be a shortage of older black tom cats. The only black cat available was one that was only half the size of the original cat and this was a 'she' and not a 'Tom.' Still it was the best that could be done at such short notice so the crewman grabbed it, returned to the boat and hoped for the best. Enter the Immigration Officer, who unluckily for the Steward, happened to be the same one who came onboard when we docked. Everything was going fine until he asked to see the cat again. The shanghaied cat was brought into the saloon. The Immigration Officer looked closely and said he thought the cat he saw initially was bigger than this one. The Chief Steward looked him straight in the face and without blinking said, "I told these fellows not to wash the cat in Persil, look what happens." He got away with it, as the Immigration Officer could not prove one way or the other if it was the same animal he had previously seen. He just laughed and stamped the papers. I wonder if he knew the original sex of the cat.

  The Pilot boarded again to steer us out and we headed for Melbourne. When we dropped the Pilot off I took up station. Once again I notified all local Radio stations, and UK radio stations that we were leaving Freemantle, bound for Melbourne. All of this was routine with me by now. Two hours on watch, two hours off watch, for a total of eight hours watch keeping. The Auto Alarm was set for the remainder of the voyage. On arrival in Melbourne I did the usual close down duties and prepared to enjoy myself ashore.

  I went ashore with the Third Mate and as we walked up Flinders Street we were invited into a live radio broadcast show. It was exciting for us to go as it was a new experience. There were singers and comedians at the show and we really enjoyed it. However, we had more important things to take care of when the show finished. We wanted to find girls. As we came to the top of Flinders Street and were about to cross the road at the traffic lights, Cyril was talking with a pronounced English accent. A Policeman who was standing at the lights turned and said, "Bloody Pommies" I immediately turned and said, "What do you mean? I'm Irish." The change was dramatic. "Gee, Pat I'm sorry. No offence. It's great to hear someone from home." He escorted us across the road waving and wishing us well. This was not the only occasion when the Irish accent got a very good reception while I was abroad. Cyril said he'd let me do all the talking from now on.

  Whilst we had been walking up Flinders Street we had not noticed two pretty girls right behind us. After some time they walked in front of us but quite close so that we nearly tripped over them. Shades of my days in Cork came to mind, when we chased girls! Eventually, we smiled at them and they responded. We chatted to them and they told us they were fascinated with our accents, so different from the Australian accent. We stayed with these two girls whilst in Melbourne and they offered to help us jump ship. In fact, besides liking the girl, I very nearly did so because I liked Melbourne so much after a week there. We spent great days swimming on the beach and in St. Kilda where there was a big amusement park. It was summer when we were there and it was very hot. This really made our stay wonderful.

Again, the day of departure came and after all the fond farewells and promises of writing and keeping in touch, the ship left. One fellow tried to jump ship but was caught and returned before we sailed on to Geelong, just a short trip from Melbourne, where we loaded the balance of our cargo of grain.

We started the long voyage home after we dropped the Pilot off and we set course for the Suez Canal. I carried out the usual Radio duties and sent messages to the Company Head Office, giving our position and any other relevant information. I sent and received messages for the crew, and often these included flowers from Interflora. I spent my off duty hours on the bridge with the Captain and Mates on duty, or in some one's cabin.

We were experiencing beautiful blue skies and calm seas, until one day I got a weather report that made the Captain look worried. He asked me to keep a careful check on all weather reports for a while. After a number of hours the weather started to change for the worse. Calm seas became rougher, and still rougher. A gale started to develop and soon we were in the middle of a storm. The ship was being seriously pounded and rolled a lot. Everything had to be battened down and shelves cleared. It was extremely difficult for me to hold my seating and eat. How the cook coped I don't know but we always had our meals. For me, I did not mind too much as I had already been through worse in the Bay of Biscay, on my last voyage. Gradually the storm abated and we got back to normal. We had lost some time and now, hopefully, we would get back on schedule again.

On arrival in Suez, we had searchlights fitted and we again joined a convoy through the Canal, only this time we were on our way home. In Port Said we ditched the searchlight, dropped the pilot, and proceeded through the Mediterranean Sea for Bari, in Italy. From here we sailed into Naples. This was a beautiful city. I spent a fair amount of time looking around here and thought of Mammy who loved the song 'Isle of Capri.' The Isle could be seen off the coast. Sorrento was also close by and it was delightful. In Naples I bought a lovely music box for Mammy. Guess what it played? 'It was in the Isle of Capri that I met her.' I also bought Mother of Pearl souvenirs. In one shop, I saw a wonderful life-like wax doll, which looked like a baby sleeping. It was so human-like that I thought it was the owner's baby. The doll's chest rose and fell like a normal child breathing. It created a great impression on me. I was going to buy it but the owner said it had to be kept in a temperature controlled environment. No way could I guarantee this. I was quite sad leaving but the trip had been long and I was ready for some time off.

I sent a number of CQ (seek you) messages to anybody who might have messages or needed assistance or even a natter, and there was always someone who wanted a chat, so I often changed channels and nattered away. Strangely, in all my time at sea, I never had the good luck to chat to any Radio Officer I knew. Soon, Gibraltar came into sight and I knew we were close to home. I say home because at that time I classed any port in the UK as home, purely because I used to start or finish a voyage from somewhere there and Gibraltar was only a few days steaming from the UK.

After passing Gibraltar, our next worry was the Bay of Biscay. We were lucky this time. It was rough but it was still reasonable weather. Now we steamed up the Irish Sea to Birkenhead. I packed my bags and wondered if I would get leave in Liverpool. I would have to wait to find out.

I tried to raise Cork Harbour Radio (EJC) but to no avail. I signed off the ship on the 19th July, 1950 after a voyage of six months. I reported to the Marconi Depot in Liverpool and handed in all the relevant paperwork. I collected my salary and I was thrilled to get a warrant to go home for a break. I looked forward to seeing my family and Molly, as her birthday was coming up on July 27th.

There was great excitement when I got home. All my family were delighted to see me and I had the usual gifts for everybody. I had bought a lovely little pocket transistor radio in a leather case. I really loved it and it was state of the art. Daddy set his eye on it when he saw how lovely it was. He was mesmerised by the clarity of reception, and the light weight of it. He constantly used it and I think now that I should have let him have it as a present but, without thinking, I took it back to sea with me. I have always regretted this decision. He was always looking to see what other stuff I had - shaving creams, lotions etc. He tried them all. Even the clothes I wore were like a magnet to him. He would always be commenting on them. Once when we were in a public toilet I was closing my zip fly when he exclaimed; "Be careful boy, you could do yourself a serious injury with that thing." I presume he meant the Zip!

On returning I had to change my name. Daddy and I were both known as Jack so daddy 'in error' had opened a few of my letters from girls. I told my friends to address any letters to John P. Lynch in future. Once I got a telegram from a girl in Liverpool addressed to Jack Lynch and daddy opened it. I had only been home for a few days at the time. It read; "Come back. Urgent must see you immediately Love, Pat." Daddy nearly had chickens. Questions came at an alarming rate. "Is everything alright Boy?" "Why does she want to see you so urgently?" "She's not in trouble is she?"

I could only smile and laugh. "Dad she just misses me, she's lonely. No big secret. No shotgun wedding, don't worry!" I continued; "Pat is the daughter of an Irish couple and is a nice friend but I have no intention of getting married or running away." I'm sure that he was shook up but things quietened down when he saw I didn't rush back to Liverpool. It's a good job he was not with me during some of my trips. I wonder how he would have coped with some of the sights and escapades I witnessed.

Molly and I met numerous times while I was home. We were not yet fully committed to each other but I certainly did not try courting any other girl at home. It would be another seven years before we tied the knot. There was not much you could do in Cork around that time so we went for walks and went to the cinema. Time flew and soon enough another telegram arrived from Marconi and I had to return to their office in Liverpool for my instructions and details of my next ship and voyage.

Chapter sixteen

Once again, after another period of waiting I was given my next assignment. This ship, the m.v. Lampania was my first tanker and was owned by the Anglo Saxon Petroleum Company AKA The Shell Co. She was built at Hawthorn Leslie Company's shipyard in Tyneside, U.K. and she was scrapped in 1961. The Radio call-sign was GKDW

I signed on at Ellesmere Port on August 3rd, 1950. Stanlow is the name of the local Oil Refinery location and a well known destination for tankers. This time it was a nine week voyage from the UK to Europe and the Caribbean.

When I returned from home leave I reported to the Liverpool office and made arrangements to stay at Atlantic House until I was assigned to a ship. This could take days before a ship became available. Atlantic House was run by priests for seamen. It was a very comfortable place and I had my own room. I made it a point to stay there every time I waited to sign on or when I signed off ships in Liverpool. Since I had many trips to Liverpool I became well known to the priests and, fortunately to the girls who used to attend the dances that were held regularly in the hall. These dances were closely supervised, and no girl was permitted to go out with any of the seaman. I used to meet a girl from there most times when I was ashore. Her name was Mary and she was tall, dark, pretty and we got on very well. However our friendship never grew into anything serious.

Once, when I was staying in Atlantic House, Her Majesty Queen Elizabeth ll and Prince Phillip visited the building. I was one of the seamen at the door, placed there as part of a welcoming guard of honour. She did her duty and then left. She just smiled as she came in and that was that. I was not really interested at the time. While I showed great respect to them I'm sure none of them even saw me, and I only stood at the door to make up numbers.

Around this time I was interested in football and supported the Everton Football team mainly because there were some Irish players in this team. Any chance I got I went to see them play at their home ground in Goodison Park- and on one occasion I watched them play their local rivals, Liverpool Football team. The atmosphere was electric and I came away exhausted having cheered and shouted throughout the game.

I had enjoyed my time in Liverpool but all good things come to an end and I got my new assignment and it was now time to sign on the Lampania and begin the first of a two part voyage. On boarding the tanker my first thought was that she was very clean and I noticed that she was very well painted, which was characteristic of all Shell Tankers, as I found out later. The smell of fuel oil was strong as I went aboard and into my cabin, to drop off my suitcase. If she was to go through the Suez Canal there is no doubt but that the British Tommies would have given a warm welcome to her. My previous trip through the Canal was evidence of their passion to be British. They would be very proud and would show in the usual cheering and waving as distinct from the ranting and raving shown to badly painted ships flying the Red Duster. With these thoughts in my mind I looked for my quarters which I found on the bridge near the Radio Room.

Next, I went to the Captain's cabin, and was greeted by a tall man who shook my hand and we spoke generally for a while before I signed articles (a nautical term for signing a contract) on the ship. I felt comfortable with him. I noticed though that he stood very straight and walked stiffly. Later I found out he wore a corset because he had a back problem. It amazes me that I cannot to this day remember one solitary crewman, officer or sailor on board that ship. This was despite the fact I was on it for nine weeks. In retrospect I think this was due to the design of tankers. Whereas on cargo ships it was possible to mix with crewmembers on deck, which was well clear of the sea and had hatches to sit on, in tankers it was not feasible to wander around the lower deck and to get from one end to the other end of the ship it was necessary to go by 'catwalk'- which is a long bridge,

*1948; Jack Lynch. 21st Birthday in
Welwyn Garden City, Herts. England*

1949; RMS Isle of Guernsey. My 1st ship.

JACK LYNCH

1949; RMS Isle of Guernsey; Crew outing in Jersey Island

1950; mv Lampania, My 4th ship

BEYOND THE SEA

1950; Cobh. Mammy (Sheila Lynch)

1950; mv President Brand, my fifth ship.

1951;
ss Orford.
Jack in uniform

1951; ss Orford. Jack in Whites

with safety hand rails - raised above deck This meant most of the time was spent on the boat deck or bridge.

I remember the first part of my voyage was to the Island of Curacao, which is part of the Netherlands, Antilles. We had left Ellesmere Port a few days after I had signed on in August 3rd 1950. As we came down the Irish Sea the weather was quite calm and we made good time. The usual sea traffic was all the way down the coasts and when we emerged into the Atlantic things looked good. We had beautiful sunshine and luck was with me when I called and got a response from EJC (Cork Harbour Radio) and I sent a message to Molly. During the trip I spent a lot of time writing letters to home and to Molly. I would have to wait until we got to Curacao before I hoped to get a letter or to post what I had written.

During the voyage across the Atlantic I remembered the pictures I saw in the cinemas about ships being torpedoed and sunk during the war. I could now feel a little bit of the sailors' fear and frustration when left in the sea to fend for themselves. Horizon to horizon there was not another ship in sight. I thought of ships in the past I had read or heard about that had got into to difficulties with no one to help them. The stories of crew members severely injured and having to listen to their mates in agony in the remote isolation of the sea really got to me. The added fear of night and stormy weather battering them made the experience more frightening than one could imagine. I tried not to dwell too long on these thoughts. As we carried on I read some Mickey Spillane books and spent a lot of time on deck watching the occasional ship pass in the opposite direction and enjoying the beautiful sunsets and sea changes from green to azure blue

In the Caribbean Sea I loved to watch the dolphins swimming and criss-crossing in front of the ship's bow. They were so graceful and fast in every movement they made. There were lots of turtles around as well, and they just seemed to float along on the sea, and under the sunny sky. It was common to see them mating, and I was told that they stayed mating for days on end and just let the sea take them where ever it wanted. It was so peaceful that I could have been tempted to emigrate to one of the islands. On later voyages in this same area I enjoyed the same beautiful sights as well as pelicans and gulls diving and swooping for fish. While I watched the birds I was particularly impressed by the pelican which always came up from under the sea with big fish thrashing in his large beak and pouch a little poem I learned sprang to mind;

> *What a wonderful bird is the Pelican,*
> *His beak can hold more than his belly can,*
> *He can hold in his beak,*
> *Enough food for a week, and,*
> *I don't know how in the hell he can.*

After a reasonable and fast crossing we arrived at our destination – the Oil Refinery in Curacao. This is the largest of the Netherlands Antilles Islands and lies between Aruba and Bonaire off the coast of northern Venezuela. When we tied up at the jetty amid shouting and lots of other noises a shore gang boarded to connect the oil pipes to our ship. In very little time this was accomplished and oil started to flow into the ship's tanks. Various officials came aboard to inspect and clear the ship and crew. Customs only visited the Captain's cabin. Amicable discussions must have taken place as we did not have visits or searches in our cabins. Likewise the immigration cleared us in record time. The Shipping Agent arrived and delivered our mail and collected what we had to send home.

While this was happening I took off to go ashore but due to time restraints I did not have any real time to browse for long. However, I did manage to get some of the usual items; Cushion covers, ornaments and chocolates. The oil soon filled our tanks and it was time to take off for home. I had now experienced the difference in time taken to load general cargo versus liquid cargo; turnaround for a tanker is much, much quicker than for a general cargo ship.

Tanks full, we collected our Pilot and prepared to leave port. Our orders were to deliver the cargo to Stanlow, in

Ellesmere Port. When we left and when the Pilot disembarked I took up station and sent my departure and destination messages to the local and U.K. Radio stations. Since I was well out of medium radio range from the U.K I went on to shortwave radio and had some difficulty raising Portishead Radio. Eventually after I tried on different frequencies I eventually got an answer and was informed that they had a message for my ship. The message was for one of the crew and I gave it to him. He was 'over the moon' as it was great news for him. Soon he was bragging that he was a dad!

With a new dad onboard and with the ship low in the sea - which overlapped the deck in rough weather - we headed back for Liverpool. Again the Irish coast came into view and the loud radio signals from the local Coast Stations were a welcome experience. I had no luck this time trying to raise Cork Harbour Radio and so we continued up the Irish Sea to pick up our pilot at Liverpool. With the correct tide conditions we tied up at Stanlow Oil Refinery and went through the normal official business. I had sent my final messages and was now ready to sign off as this trip was finished. First of all I collected my mail which had been delivered and sat back to enjoy reading the latest news from all at home. It was September 23rd 1950 when I signed off but Marconi requested that I re-sign for the next voyage and I was glad to do so as it was to be a short voyage to Europe and back. I re-signed on the same day September 23rd 1950. I posted more letters home and I went ashore for some exercise and browsing around as there was no home leave this trip. In the meantime there was more ship's business to be conducted before we could begin the next part of our voyage.

The Steward arranged for food and other essentials to be delivered and a few new crewmembers were signed on to replace some who had left, including the new dad. We took bunkers (fuel) onboard and when the tide was suitable and everything was ready the Pilot came aboard and we left Liverpool to enter the Irish Sea again. We came entered the English Channel, headed for our next destination which was Rotterdam, Holland. This is a large port with a lot of Industries, including Petrol Chemical

Refinery and is used by ships of all nationalities. With the aid of a Pilot we tied up and went through the official inspections and safety checks. The dockers here were experts and had us connected, loaded and disconnected in record time. Soon we had the Pilot back onboard and we headed out into the North Sea bound for Stanlow, in Ellesmere Port. The ship sailed into the English Channel and back into the Irish Sea. This time the weather was foul and the decks, which were low in the water, became almost submerged. It was frightening, and not having previously been on tankers at first it worried me a lot. The only way to move on the ship was by the catwalk. However, it was only a phase and I soon got the feel for it.

Again we arrived and picked up the Pilot at Liverpool and went into Ellesmere Port where we tied up at Stanlow Oil Refinery where we discharged our cargo and headed downriver to Birkenhead, Liverpool. I signed off on October 11th 1950. I shook hands with the Captain. In the Marconi office I had to report in and await a new ship. Fortunately, I got some more leave and went home for a few days and enjoyed the usual excitement and relaxation with my family and Molly prior to joining my next ship destined for I knew not where.

Chapter seventeen

My fifth voyage took place on another tanker, the m.v. President Brand. Her original name was Egg Harbour and in 1948 she was renamed President Brand. As she was registered in Capetown, South Africa she flew the South African flag.

I signed on this ship at Ellesmere Port, Lancashire on October 27th, 1950 for what was to be my fourth deep sea voyage to the Persian Gulf.

This ship was a WWII T2 Tanker and the biggest ship I was to sail on so far. She looked like a monster to me when I first saw her though by today's standards she would be classed as medium. They were built in the USA, in the 1940's during the war, and were the workhorses of the convoys, shipping oil to the UK and Russia. President Brand was built in April 1943. The lifespan lasted well into the 1960's.

My cabin was a good size with its own toilet and shower facilities. There was a comfortable bunk with good wardrobe space. Again the cabin was located close to the Radio Room and near the ship's bridge. In the Radio Room the RCA equipment was of a type I had not experienced so far. It looked classy and it did not take me long to figure out the whole setup. I did notice a shortage of some spares for radio replacement parts and put in a requisition for these and they were delivered before we sailed. All else appeared fine so I waited to go on watch when we weighed anchor and dropped the Pilot as we left Liverpool.

I was surprised to see the South African flag flying but the owners must have had a contract with Marconi for the supply of Radio Officers otherwise I would not have been aboard. It was a nice trip, which took us down through the calm of the Bay of Biscay, onto Gibraltar, and then into the Mediterranean. So far the weather was quite good all the way but now we started to get some strong sunshine and it was lovely to see the ship cutting through the calm blue sea. There were various glimpses of land on each side as we headed for Port Said, to go through the Suez Canal.

As we progressed on our voyage I was beginning to get quite annoyed with one ship that was out of sight but close enough to almost blow my brains out when the Radio Officer opened up his transmitter. This happened a few times daily, when he kept asking if anybody had anything for him. I got to know a Radio Officer's "fist" (his style of sending Morse) and the tone of his radio. This fellow's call sign indicated that it was a Greek registered ship. I was wearing headphones when this guy opened up and by now I was fed up of him blasting my ears and so I committed my second cardinal sin at sea. I waited for him to come on air as usual, and when he sent his first message; "CQ, CQ, CQ," I immediately replied on full power; "No, F*** off and shut up." Imagine my shock, and shame, when the guy came back in perfect English; "You are a perfect gentleman sir."

I had assumed the guy was Greek but there were plenty of freelance English speaking Radio Officers manning ships of all nations. The guy could even have been Irish! Never again did I use foul language on the air. I was lucky that nobody recognised my "fist" or I would have been in serious trouble. Anyway, despite that big mistake things went smoothly after that and the fellow didn't use his transmitter very much - just once a day, and there was glorious sunshine and coastline to admire and enjoy until we arrived at our first port.

In Port Said Since we were not berthed so no one could come aboard or go ashore. The Agents who represented the Owners and took care of the Captain's requirements arrived and went to the Captain and soon after that we had searchlight fitted and joined the convoy, heading south to Suez. There was the usual mix of various types of ships in this convoy, including two American warships. The bumboats, of course, were in full swing, trailing alongside our convoy. Anything you wanted you could buy without having to go ashore. Eventually we passed through the Canal and the British troops as usual were shouting abuse at the British Flag ships, which did not come up to their decorative expectations. We waited at the Bitter Lakes for the other convoy going north to pass before we continued into the Red Sea. Here, we had Egypt, Sudan, Eritrea and Ethiopia on our starboard side whilst on our port we passed Saudi Arabia, and Aden in the Yemen. Now it was getting quite hot, and it was also humid. From Aden we went on into the Arabian Sea, hugging the coasts of Yemen and Oman. Next, we came to the Gulf of Oman, passed Dubai, the United Arab Emirates, and the coast of Saudi Arabia. On our starboard side all the way up stretched the coast of Iran. Qatar and Kuwait were left behind before we arrived at our destination Basra, in Iraq.

During the trip into the Gulf the humidity was very high, and I felt muggy and clammy. There was also the smells from the water, which combined with the humidity was not very pleasant. One thing that stands out in my memory is the phosphorescence glow in the sea, as the ship sailed on through the night; it was like a magical sprinkling of lights across the water. Many a time I wondered how the marine life lived with the phosphorescence. To me it was awe inspiring, to see this glow at the bow of the ship. It was difficult to sleep at night, as there seemed to be no air and fans did not help. I longed to just be back in a warm dry atmosphere, where sleep was easy to come by.

When I first heard of our destination was to be Iraq, my mind again returned to the movies I had watched at home. I thought of 'Arabian Nights', 'The Thief of Baghdad', and 'Sinbad the Sailor' but was very disappointed when we arrived. There were no bright coloured garments and nobody wearing swords. No magic carpets! In fact all I saw were ordinary peasants going about their business. There were quite a number of army fellows and their dress was very untidy. They looked ill kept and slovenly.

We did not see too much as we loaded up our cargo of oil in a short time. However we did get to see the famous, or should I say infamous, 'Bull Ring' in Basra. This was a well-known place where many of the local men and visitors went for pleasure. At the time I went there; believe me I had no idea what the Bull Ring was. My idea of what it was similar to another American Captain's I sailed with, who thought Piccadilly Circus in London was in fact a real circus with animals, and he bought a camera to take home photos of these animals! The Bull Ring, to me, must have something to do with Bull fighting, I presumed. How wrong I was! It was one huge brothel area as I soon discovered. As a number of us approached a walled area and were going through an opening, we were accosted by police, who wanted money before they would let us into the 'Ring.' Having tipped them we went towards three or four story blocks where men were entering. We followed, and the inside belied the dusty, sandy coloured outer walls. The floor was apparently made of marble. In the centre, sitting in a circle on the floor, were a group of men playing some sort of game. It looked like draughts, or something similar.

The inner area was a quadrangle built around the square floor. As we climbed the stairs and walked along each balcony overlooking the floor, there were lots of small rooms with all kinds and shapes of females inside, who tried to entice us to go with them. They wore the scantiest of clothing and many were nude and we saw lots who had tattoos also. Gradually we got through the gauntlet, and proceeded to the top floor to see what was different. Nothing changed, except we could hear grunting, and groaning from behind closed doors.

We were now on the top floor, where we found a loo, which was used by some of the lads. My mind was elsewhere. Discreetly, I unscrewed the light bulb above me. I looked over the balcony, and could see that the men were still playing their game. "Here goes, get ready to move" I said to the lads.

Then I dropped the bulb over the balcony, and it landed very close to the men with a bang as it exploded, shattering into pieces beside them. The men immediately jumped up and scattered. We did not wait to see what else happened. When we got to the gate we sauntered back to the ship, hoping we would not be picked up for disturbing the peace.

When we left Basra, we proceeded down the Gulf and called into Ras Tanura, in Saudi Arabia where we loaded some more oil. Then it was onto to the Suez Canal where the pilot came aboard and we had the usual searchlight fitted. When the convoy started to move north we took our place between a passenger ship and another tanker, which was behind us. All was well but there was a following wind, which was quite strong. We reached the Bitter Lakes and proceeded to move north, towards Port Said. After about a half hour our ship suddenly started to belch smoke from the funnel, and the pilot shouted to the Captain to get the situation under control. Sparks from the funnel lit up the night sky, and the following wind blew these sparks right across the tanks containing the oil. The pilot kept screaming at the Captain and he looked as if he was ready to jump overboard. The ship behind us signalled she was slowing down to get out of the way. The ship in front could do nothing, as it had no room to manoeuvre due to the narrow canal. I prayed, and I'm sure I was not alone, as the Chief Engineer did all he could do to control the problem. Being night-time, the sparks looked ominous and frightening. There could have been a deadly explosion if it continued much longer. After about twenty minutes the smoke died down and there were no more sparks. Everybody sighed with relief. We hoped to continue as normal but the pilot had to report the incident. We discarded the searchlights, avoided the bumboats, and set off back to the UK with our oil.

During this particular voyage I'd found that the food onboard was terrible. There were weevils in the corn flakes and cockroaches all around the place. When I complained about the weevils and cockroaches the boys laughed and said "it's all good fresh protein."

Generally, the weather was good and did not create any problems. We eventually arrived back in Heysham, Lancashire where I signed off and reported to the Liverpool office. It was December 1950 and I was given some leave to return home for Christmas. We had a wonderful Christmas and for the first time at home I tasted turkey. Though I enjoyed it I longed for the goose like we had when we were growing up. The days brought a lot of childhood memories back to me and we all watched TV in black and white. It was wonderful, it was home and it was just like the old days and I enjoyed every minute of it. The one thing that I missed was my pals that I grew up with. They were all now gone from the town and there were very few that I could claim to be real friends. On my previous visits I was quite content to spend time with Molly and my family but this time it was different. Maybe I had seen so much abroad that I just wanted to return to the old days! In Cork I met Molly on a number of occasions and we had a lovely time walking out the Mardyke to Fitzgerald Park or going to the cinema. Strangely, we never went dancing as I used to go to the Arcadia in the earlier days with the lads but Molly did not like this hall and she usually went to Glenbrook, which she thought was more sophisticated!

When the inevitable telegram arrived again I reported back to Liverpool on the 16th January, 1951 to wait for my next ship. I checked in at Atlantic House and got ready for my next assignment.

Chapter eighteen

I signed on to my sixth ship, the s.s. Orford on January 1st, 1951 in Manchester. This was a Canadian Type Liberty (or Park) ship, registered in London, and Owned by Ship's Finance & Management Board. The Master of ship was Captain Pinkney and Mr. Bond was the Chief Engineer. Mr. Contaratos was our Second Engineer and the Chief Steward was Mr. Moxon. The remainder of the crew were British and German. Most of the Radio equipment on this ship was Canadian built.

This was to be my fifth deep sea voyage and was to Argentina, Uruguay, Germany, France and back to the UK. From the Liverpool office I was told to join my next ship in Manchester and was given a travel warrant to go by train. At the station in Manchester I hailed a taxi and requested the driver to take me to the ship. I arrived safely and lugged my baggage up the gangway to my cabin on the bridge. My first call was to the Captain's cabin where I introduced myself and signed articles. The Captain appeared to be quite a nice person and I also met the Chief Engineer. I returned to my cabin, which had the usual bunk, toilet facilities and lounging chair, and unpacked and connected my personal radio; a lovely shortwave Philips radio. This was to be a great investment during my many voyages. It was great company, even for a Radio Officer who had no shortage of radios! The rest of my clothes, books and writing materials were now put away for the trip and I settled in for a few days to await departure.

The ship was not ready to sail due to a crew shortage. The Chief Engineer needed a fourth Engineer and asked me if I knew anybody who would take the job, as the company were not having any success trying to recruit a suitable candidate. I immediately thought of Joe Murphy and felt he could do it. I rang the Guards' Barracks in Cobh and gave them my number in Manchester. I then asked them to contact Joe and get him to call me back, which he did. I spoke to Joe who jumped at the chance. Then I told the Chief and gave him Joe's home address.

Joe was sent for by the company and told to report to Bevis Mark House, London, for a test. I had no idea what was going on for a few days until the Chief Engineer came and thanked me because Joe was on his way up from London having been accepted for the job.

I kept an eye out for Joe and eventually a taxi arrived at the jetty where we were berthed and out popped Joe with his bag. For the hell of it I watched him struggle with the bag and eventually he came aboard puffing and panting, as I had done on a number of occasions. He came round after a short while and I asked him what kept him. He gave out a mouthful and said that when he got to Bevis Mark House nobody seemed sure who he was or why he was there. He explained he was told to report there for interview to check his suitability as a Marine Engineer. He was then taken to a room and was given a paper with questions to be answered. He nearly choked. The questions were way above his head and he could not do any of them. His mind went blank and he thought that his trip was a failure. Once he told them he was applying for a fourth engineer's position and not that of a Chief Marine Engineer however, things changed. They had given poor Joe a Chief Engineer's examination paper by mistake. The new paper presented no problems and it was with deep relief that poor Joe got his train to Manchester. We chatted for a long time and Joe was over the moon to be starting a life at sea. His brother Tim, who had been in my class at school, was also at sea as a Marine Engineer with Irish Shipping. Both of these boys rose to Chief Engineer rank. In fact, Joe was promoted to Second Engineer on this ship before the voyage finished.

After a few days many incidences had occurred amongst the crew and I felt this voyage was going to be too long, even though at the time I was not aware it was going to be such a long voyage. When I signed on in Manchester in January of 1951 most of the crew were already on board. One of the crewmen consisted of an ex-Irish guardsman from the north of Ireland, and there was an ex-Scottish Black Watch soldier from Glasgow. He was not the full shilling and suffered from shell shock. The ex-Guardsman stood six feet two inches tall and had all his teeth filed into points. During the voyage he told me that he had this done so that he could better protect himself from the Germans during the war. If an arm came round his neck he could just chew the arm to pieces. He frightened and intrigued us in equal measure. Some of the crew came from Wales (Taffies) and some from the northeast of the country (Geordies) and I think there were two from Liverpool (Scousers). Anyway, it was a right mix and most were hard drinkers. Nothing unusual with the crew so far but later things would change for the worse.

We left Manchester for Buenos Aires in Argentina, proceeding down the Irish Sea and out into the Atlantic Ocean. I followed all the usual Radio routines and tried to pick up EJC (Cork Harbour Radio) as we passed the south Irish coast. They may have heard me calling but there was no response so I gave up and sent my last departing message to Valencia Radio in Co. Kerry, advising them of our destination and where we had come from.

About a week into the trip the crew complained that they were not getting their full rations of food. It was common practice on ships, when the sailors signed on articles, that an agreed amount of each type of food was the right of all who sailed on the ship. It was called 'the whack.' The crew got very frustrated and eventually each asked for his 'whack,' which meant that the Chief Steward had to individually weigh each and every bit or allocation that was due to each seaman. This had to be done in the presence of each seaman involved, and they would carefully watch the scales so that not a hint of overweight, or underweight, took place. This nearly drove the Chief Steward mad. Imagine trying to weigh butter and other foods precisely. Not only that but the Steward had his son on board as a cabin boy and the sailors took it out on him too. They used to urinate over him in his bunk and this was really disgusting. The Captain intervened and cautioned all the crew involved, threatening them with imprisonment for mutiny. The conflict stopped but the damage had been done and there was an uneasy peace onboard. Some of the crew were dismissed when we returned to Europe for this. You could never be sure how a group of strangers, sometimes with different cultures, would react when closely confined on a ship for a period of weeks. When drink was involved the chances were very high that things could go very wrong and I think this is what caused most of our problems on this ship. In all my time at sea I cannot remember any drugs being used except alcohol and tobacco. Amidst all the fighting and arguments there was always the occasional funny and joyous occasion to keep you sane.

The man who keeps the deck crew in order and the ship in good condition has similar responsibilities to a foreman and is called the Bosun. His wife was expecting a baby. He kept talking about it and asking me if any message was sent for him to confirm that the baby had been born. I kept telling him that there was no message for him, and that I would tell him when the happy event took place. Eventually, the telegram arrived and I gave it to him. He was over the moon, and confided that he had been sitting on the loo when the event took place and that he suffered 'Symphony pains' at the time. I nearly burst a gasket laughing! He kept telling everyone later about these 'Symphony pains' and I did not have the courage to tell him that the pains were supposed to be 'Sympathy pains.' Nobody else did either. He was the same fellow who one night tied a piece of string to a young boy's penis while he was asleep and then shouted, "Fire, Fire." The poor lad jumped out of his bunk and nearly did a serious injury to himself. The Bosun and some of the crew found it hilarious. The poor lad was the Chief Steward's son and it was his first voyage, and as I said the crew had a vendetta against the Chief Steward.

The voyage continued to South America and we hit some mountainous seas on the way. The ship was tossed and pounded as we proceeded. During this period I nearly went crazy with the short-wave receiver. It had plug in coils, as distinct from fixed coils. To change frequency it was necessary to plug in the respective coil and then use two dials to tune the set to the required frequency. This was a cumbersome and touchy operation at the best of times in calm seas but during swells and pounding seas it was an outright bitch. The coils kept jumping out of their holders, and I had to find them and retune the set all over again. This was a nightmare when I tried to send or receive messages on short wave. I had to resend and ask for repeat transmissions a couple of times during the trip. Fortunately, the storms were not too frequent on this voyage and I coped without going bonkers, as this was a very long trip and it was my luck to get this damn receiver. Never again did I have the misfortune to meet up with such obsolete junk radio.

After a few days we got back into clear weather and gradually approached the coast of Brazil. As we passed we could see some of the coast quite clearly. I will never forget the sight of horses' heads floating in the water, just off shore. The Captain said that the horses were slaughtered for food, and the heads were put to sea for sharks or other predators to eat them.

Down this coast it was almost impossible to read Morse messages, due to the loud and constant atmospherics. The automatic alarm was activated by these atmospherics, and I had to get up every night for false alarms. Eventually, I left the alarm off and reported this in my logbook. About two days later the radio signals got back to normal and I reset the alarm during off watch periods.

Throughout this voyage I met Joe quite often and we chatted away, getting on well. Joe used to get himself into all kinds of harmless trouble. Being the bragger that he was many of the lads were out to get him. We had a Danish steward and Joe used to class him as being a bit 'Queer' and used to bend down in front of him to tease him. One day Joe was in the shower and along came the Dane, who spotted Joe and shouted; "Now I've got you." He went in after Joe, who was covered in soap. Just in time Joe saw him and managed to squirm free, covered in soap and naked, he ran screaming down the corridor. Joe never got over this and did not tempt providence again.

When we were crossing the Equator, Joe was the target for the crewmembers to give him a dose of 'Neptune's welcome.' This could be anything that was discomforting, like being tied up, stripped, and hosed down. Some more dangerous tortures were also carried out. As we approached the target area the crew lined up to get their hands on Joe. He must have had a sense of the imminent danger because he would not come out onto the deck. Patience was eventually rewarded. The sailors left the main deck and hid in the boat deck overlooking Joe's exit point to the main deck. I stayed on the main deck to allay his suspicions and as he came out slowly, looking everywhere before venturing out. It took all my will power not to laugh. Joe stepped out onto the main deck and down came three buckets of rubbish, slops and sludge, which would normally be dumped overboard. Joe got the whole lot of leftovers, peelings, gravy, and whatever could and would not be eaten. He was a sorry sight, and took off quickly in case he got more. He gave out stink to me about our friendship, or lack of it. On my first trip to Australia on the Tower Hill I too had to cross the Equator for the first time I did not have the ignominy of suffering Neptune's baptism. It's just as well that he had not known this and I didn't tell him.

Joe was a glutton for sweet things, and always wanted more deserts. One day, the steward challenged him that he could not eat a full tin of pears. Joe said he could so a crowd gathered to see if Joe could finish the tin of fruit. It was not a level playing pitch by any means and little was Joe to know he was being played himself. The ship was rolling in a moderate swell. Joe had already eaten his dinner when the steward came out with a very large catering size tin of pears in juice. He handed it to Joe and gave him a spoon. Joe complained that he meant a normal size tin of pears but this cut no ice. We all watched as Joe quickly got through about half the contents before he started struggling. Gradually, his face changed colour, as he pushed more and more pears into his mouth. Suddenly, Joe was hanging over the side of the ship, and up came pears, juice and a lot of dinner. Poor Joe was learning the hard way.

Eventually, we reached the River Plate, picked up the pilot and proceeded up to Buenos Aires. I closed down the Radio room and was technically free until we dropped the pilot when we were leaving. On the way upstream we had Argentina on one side and Uruguay on the other. The remains of the German pocket battleship 'Graff Von Spee', scuttled during WW2 was just at the mouth of Montevideo Harbour.

When we were tied up I went into the city. It was beautiful and there were lots of pictures of the Dictator Juan Peron and his lovely wife Eva. Everywhere I went these pictures were displayed. When I first saw these large posters outside the cinema I remarked to one of the ship's officers;

"Is that the tyrant I've heard about?" Pandemonium ensued! A hand was quickly put over my mouth and a voice whispered;

"Do you want to end up in jail and probably never be seen or heard of again? This place is full of vigilantes." After that I learned my lesson and kept my trap shut.

One evening, most of the crew and the Captain went to a bar that was owned by a well-known local tenor. We all sat around enjoying his singing and the attention of the rest of the male staff who could not do enough to please us. Most of them spoke very good English. I didn't cop on at the time that there were no women serving drinks and on reflection I can't remember any females in the bar.

I soon got an inkling of what was going on when one of the barmen asked me if I would like to go and see a movie. "Sure, I'd love to." I replied, innocently.

We met the following day and headed into the cinema. We took our seats and while the cartoon was still showing I felt a hand touch my knee. I got a funny feeling things were not quite right. While I was wondering about this Johnny, my friend from the bar, then put his hand on my knee, and left it there for about three seconds, before I pushed it away. I was feeling really uncomfortable by now and was considering whether I should get up and go, when his hand made a grab at my 'Jewels.' I grabbed his wrist, twisted it and whispered; "Johnny, stop. Get up and go or I will clobber you"

He got up without a word and left the cinema. I sat there, not seeing anything on the screen or even hearing a word. I just thought, 'this is Argentina, a dictatorship, and if Johnny accuses me of anything, who will believe me?' I waited for what seemed hours and vigilantes did not arrive so I got up and headed back to the ship. All the way back I was uneasy and uncomfortable. No more going ashore in Buenos Aires for me.

Back on the ship there were various stories coming out about this bar. They were all Gay in there, even the owner. Joe told me that one of the attendants asked him to go to bed with him, and promised him all kinds of presents. Joe, being the teaser that he was, promised this guy a great time but like me and many more, he too stayed on board. The Captain was propositioned as well as others who went to the bar. One seaman told me that he got drunk and woke up in a strange bed with two fellows next him stark naked. He could not remember what he did or what happened.

We weren't sorry to leave Buenos Aires and head down the coast to Bahia Blanca, in Argentina. It did not take too long to get there and suffice to say I was not taking any more chances in Argentina, and was glad when we dropped the pilot and headed out to sea.

We had orders to proceed to Hamburg, Germany. The trip, from what I remember, was OK. We progressed up through the English Channel and into the North Sea, and came to the river Elbe, where we picked up the pilot. The river was extremely busy with lots of ships tied up at the docks, and others moving in and out of the Port. Eventually we berthed alongside a jetty near the centre of Hamburg so it was a nice walk into the main part of the city. There was not too much war damage visible from where we were but I did see some later whilst ashore.

As we had returned to 'Home waters' the Captain decided to get rid of the trouble making crew, who caused so much havoc. This had been pre-planned by the company, as there was a replacement German crew waiting in Hamburg. I was not privy to this decision as I had not sent any messages from the Captain pertaining to the crew's behaviour. I presume that the Captain had phoned Head Office from Buenos Aires and advised them of the situation onboard. This crew change was to cause more problems later. Most of the English crew were sent home, as were a few deck officers who signed off by choice.

We were all anxious to get ashore and sample the life there. We were not disappointed. St. Pauli was soon discovered. This is where life was taking place. The neon lights were ablaze, and there was music from all the pubs and clubs. Beautiful girls were mingling with sailors of all nationalities, and drink was flowing like the river Elbe. American servicemen were there in abundance, as were the American Military police who occasionally raided a building to check if the prostitutes were carrying their medical certificates to show they had been recently checked for VD.

As I sat with the lads drinking my soft drink and they were getting jolly on lager, I was watching couples going out or upstairs, only to return sometime later looking the worse for wear. My friends were themselves getting restless and decided they were going to get partners for the night. All this time there were girls coming up to our table, asking if we needed company but so far none had been invited to sit down. One by one the lads surrendered and selected a girl to bed with them until I was left alone without company. The sweat was beginning to roll down my back as the lads encouraged me to 'go for it.' In my mind I could see the Jesuits and Redemptorists pointing fingers at me, thumping the pulpit in St. Colman's, warning me of the 'Fires of Hell' if I as much as thought about it. Well I had, and was halfway to Hell now. But if I fought the temptation I could still be saved, and also avoid a possible infection, which was a dread I could not contemplate.

"Come on Sparks, what are you going to do?" A voice asked.

Peer pressure! The girls were lovely and my spirit was weakening as the clock ticked. A girl slipped in at my side and took my hand in hers, whilst she smiled up at me. This made up my mind for me. My knees started to wobble, and I had no resistance left.

An Argument.
I've oft been told by learned friars,
That wishing and the crime are one,
And Heaven punishes desires
As much as if the deed were done
If wishing damns us, you and I
Are damned to all our heart's content;
Come, then, at least we may enjoy
Some pleasure for our punishment!

Thomas Moore, Poet. (1795-1852)

All the sermons I had heard during the missions were forgotten. I followed the three lads, with their partners, and this girl through the bar and up the stairs where we met more fellows and girls on their way down. My mind was in turmoil. Should I run and get the hell out of here or should I be a man and face up to the challenge? If I backed out now the whole crew would know and I would be the laughing stock on board. There was nothing for it but to go through with it, if I could. Doubt began creeping in again. What if I could not perform due to tension? Why did I get myself into a fix like this? Whilst these thoughts were flowing in my mind we arrived at a door on the second floor? I did not know where I was, by now some of the lads had already gone to their rooms. When I saw the door to her room I got a shock. The building must have suffered war damage because for the first time I noticed all the doors were unpainted planks of wood roughly nailed to cross pieces to hold them together. They were hung on rough wooden frames, nailed to the wall with plenty of unplanned ventilation and light holes. To crown it all the only lock was a padlock and bolt in the center of the door. I mulled over this as I entered the room, and immediately noticed the view from the window was beautiful, overlooking the river and city. Lights twinkled on the river below and the streets were aglow with coloured lighting. I was to have plenty of time to admire this scenery as the night passed.

The girl asked me for her money before she became 'friendly'. After I paid up she smiled sweetly, and said she had to go down to the bar to tell her friend she would be occupied for the night. She told me to make myself comfortable. I noticed that the bed was large and clean. There was an armchair and various other pieces of old but good furniture. Whilst I was absorbing this, the door closed and I heard the bolt being drawn and the lock clicked into place. Why was I being locked in? I called her but she ignored my call. Thoughts of all descriptions were flowing in my head and I can assure you they were not of the Jesuits or Redemptorists.

"The bitch, who did she think she was? She was going out to spend my money without earning it. Was I set up for a mugging?" Thoughts flooded my mind. Gradually, as the hours passed, I kept looking out the window trying to get a glimpse of her but with no luck. I had tried the strength of the door, and felt I could knock it down if I had to. Instead, I shouted, hoping that one of the lads would hear me. After some time they came to the door and asked what was wrong. I told them and asked if they could open the lock. One got a knife and tried to unscrew the bolt screws without success. I told them to stand back, that I was going to knock the door down. They asked me to hold on, as they wanted to get dressed to get out when the commotion started. By this time there were a lot of girls and fellows on the stairs watching, and waiting, to see what was happening. The lads came back fully dressed, and I stood back and kicked the door and frame very hard. It started to give and on the second kick the entire door and frame came away from the wall, tumbling flat on the landing. There were screams from the girls as I emerged in a flaming temper. I told them to tell the "who**" that I would be back for my money, and that she could pay for the damage to the door." If I'm honest I was glad to have an out and I was very nervous.

The lads and I left and went back to the ship and I vowed to go back and get my money the following day. It was five a.m. By this time I was more tired than if I had spent the entire night with 'yer one.' Thankfully, my virginity was intact!"

As promised, the following day I went back to the bar with some of the lads and asked where the girl was. Nobody wanted to spill on her so I went to the guy behind the bar and relayed the story to him and asked him to pay me and get the money from her. He shook his head and just then I saw an American patrol passing by and told the barman I was going to report the incident to them. He immediately panicked and gave me my money. I now felt much better as I told him she charged more than she did, and so I got a few bob more and hopefully she had to pay for the door too!

I spent many happy days ashore admiring the city and it was great to see people enjoying the beer gardens. I was to return to Hamburg again on two more of my ships; the Albert G. Brown and the Esso Birmingham but nothing as exciting or embarrassing as this happened again.

From Hamburg we called to Antwerp, in Belgium. I can only remember going ashore once with some of the lads. We went to a bar on the docks and there were the usual call girls. They were looking the worse for wear. The smell of the place was foul and I got out very fast, never to return. Good riddance!

Next we went to Bordeaux, in the Bay of Biscay. I enjoyed this city and spent a few days looking around. It was fortunate that the Bay was in a tranquil mood for us and we had lovely weather and calm seas. From Bordeaux we again headed into the Bay of Biscay for the USA.

Eventually, we arrived in Philadelphia and loaded our cargo, coal, on board. Loading coal is slow, as the usual cranes and grabs were used and the cranes had to be manipulated carefully. We had some damage done to the ladders in the holds by the grabs smashing into them. There was also the coal dust flying around the place. However, the time taken to load the ship gave us ample chance to go ashore and see the city. I felt at home there and it was full of life. As usual the low lives were around, touting for business. They seemed to sniff naval men from a mile off.

It was here that the Chief Stewart Moxon and I went ashore to visit the local Radio station, WJMJ (Jesus, Mary, Joseph) that we regularly listened to on the ship. I got quite a shock when I was waiting to meet the disc jockey. When the song was finished he announced;

"Now friends, we have a full blooded Irishman here today, and I am going to have a chat with him about the 'old country.' My stomach churned and I waited for the call not realising that another Irishman had just returned to Philadelphia, after a holiday in Ireland. Thank God! We got a tour later without meeting the DJ.

The crewmembers were invited to a get together by some ex Pats. These were people who ran the Flying Angel hostel for seamen, and after prayers, hymns and something to eat we were given navy blue hand knitted woollen sweaters as gifts. It was most enjoyable. This was my first of many trips to the States and the scale of everything was enormous. The traffic, drug stores and clothes shops were something to behold.

It was here that I was to come across the American style of Customs, Immigration and medical checks, which were carried out on all crewmen. The officers were generally courteous and could be quite humorous at times. The medical checks were generally for vaccination and venereal disease.

We left port and returned back across the Atlantic towards Europe, with a load of coal bound for Rouen, France. Crewmembers were generally carrying out their duties well but there were some underlying tensions that were coming to the top. Some crew members had vendettas against each other, and drink brought these to a head in port. I witnessed some serious episodes later on but with no drink being available onboard the ship a relative calm prevailed.

Personally, I never had the inclination to become involved in fisticuffs. However, I almost had to become involved with the Second Mate on one occasion in Port. He was drunk and bust into my cabin at one a.m. I awoke to find him standing, or rather leaning, against the cabin door with a large dagger in his hand. He was calling me all kinds of names and addressed me as Finch. I can tell you I stayed perfectly still, sitting up in my bunk, but was ready to throw bedclothes over him if he came towards me.

"I'll kill you Finch, for getting me caught by immigration in Australia," he shouted.

By this time the commotion had alerted the Captain and Chief Mate, who promptly calmed him down and handcuffed him. He apparently mistook me for another Radio Officer whom he sailed with and he said that this fellow had informed immigration in Australia that he had jumped ship. He had been arrested and deported, and had a personal grudge against this Radio Officer. I was very relieved that he had been stopped before he could do any damage or cause injury, but in his condition I think I could have coped.

Later on during the trip, whilst I was asleep, my Auto Alarm went off to alert me of a possible distress message. I immediately went to the Bridge and into the radio room, only to find out it was a false alarm. The man at the wheel looked at me, winked and laughed, whilst my dagger wielding Second Mate was on watch, apparently not taking any notice. Later, I spoke to the crewman, who was on the wheel and he said he never saw anything so funny as what happened that night. The Mate apparently went into the Radio room to borrow a screwdriver, and as he opened the drawer the alarm bells went off. He got the fright of his life and rushed back to the bridge in a panic, thinking I had set a trap for him. I laughed till my sides ached. When the story got out most of the crew thought it was hilarious. To say I was happy to see the end of this Mate would be an understatement. He was one to go at our next port of call.

Eventually, we picked up our pilot at the mouth of the Seine and proceeded upriver to Rouen, where we tied up alongside the wharf. Again, I now had a first experience with a 'Boal' which is a tidal wave which runs up and down a river each day. When this occurred the ship's tie ropes, both fore and aft, were loosened twice daily as the wave lifted the ship a considerable amount and the ropes could have snapped under the strain. Whenever a ship was tied up the dock people used to stand by, and their warnings were shouted along the river, advising that the Boal was approaching. Dockers immediately sprang into action to release, and retie the ropes. The ship lifted as the Boal passed and it was similar to being in dry-dock as it flooded. I went ashore a number of times and stood at the memorial to Joan of Arc, where she was burned at the stake. For some strange reason I felt at peace there.

  When a number of us went into a bar, even though I still did not take alcoholic drinks, I again got another shock at the toilet facilities. I went into what was in my opinion the 'Gents' and saw a number of fellows lined up against the wall, in what looked like a corridor. I did likewise, and was merrily piddling when I heard female voices right behind me. I wet the front of my pants in the rush to hide myself. As I looked around there were three young girls passing behind me and going into the Ladies, which was at the end of the Gents. The ladies had to pass through the gents to get to their loo. I seemed to be the only one who was shocked but again this was a first. I soon got into the swing of things, or so I thought.

  Later at night, I was in a bar with some of the crew and there was a singsong going on, with some of the local girls. In came a woman in her fifties selling flowers. She had a basket of flowers in her arm, and had on a long skirt to her ankles. Over her shoulders was a shawl. Immediately, some of the girls started pushing her to a table and made her get up on this table and dance. I got the impression this was part of the normal evening's entertainment. She entered into spirit of it all, and after a drink started tapping away on the table. After some time the girls shouted in French to her. She grinned and started to twirl lifting her skirt to her waist. Another first for me; she wore no underwear of any description under the skirt.

  "Bless me father, for I have sinned…" Everybody got into the spirit of the night and the woman was drunk out of her mind when she left the bar. On the way back to the ship with the Second and Third Engineers, we were unsure of our way to the dock, and the poor Third Engineer had the bad luck to ask a Frenchman where our dock was. The Frenchman just turned and socked him straight on the nose, at the same time saying in perfect English "F***ing English." We eventually got back to the ship but that was not to be the end of the fisticuffs.

There were a number of fights amongst the crew as old scores were settled. One of the most serious was when an older seaman had too much to drink and got into an argument with a younger member of the crew. The young sailor was knocked to the steel deck and was unconscious from the impact. The older one then lifted the young man's legs and pulled him across the deck. The back of the young fellow's head was badly skinned from the banging against steel plates and rivets of the deck. This happened at night, and it was a while before crew members could go to the aid of the young man. He recovered after some time. The other guy was one of a number who were fired.

In Rouen there was a kind of a market where many stalls were selling fruit, vegetables, bread etc. One stall caught my eye. This was a shooting gallery, and the prizes were bottles of Bordeaux wine. I was with a few others, and I decided to show off my shooting ability. It took me one go to find that the rifle sight was slightly bent, and when I compensated for this I was knocking the little ducks down with no bother. Bottle after bottle was handed over as I kept dropping the toy ducks. Each of the crew members got two bottles each and the poor gallery owner was thrilled to see us go. He made no money and I did not drink any of the wine.

When we left Rouen we headed for the USA again to load more coal and return to Brest in France. During the outward and return voyages nothing of consequence happened and the crew appeared to behave. This was the calm before the storm as far as the crew was concerned and we really ran into trouble with the German versus the British crews when we docked in Brest. I got to know the German Second Mate and Third Mate quite well during the trip. The Second Mate was in his late twenties and did not have the Germanic features I associated with this race. He was slim, quiet and had black bushy eyebrows which met at the center of his forehead. His English was quite good but he did not talk much about himself. The Third Mate, on the other hand, was younger, stout, fair-haired and jolly. He used to tell me of his experiences in the German navy during the war. He was an 'E-boat' commander, based at Cherbourg, where he used to attack British convoys and ships in the Channel, until the RAF began to bomb their naval bases, causing them to move out of the channel and into Brest, and eventually into the Bay of Biscay. He said the RAF gave them no respite.

    A lot of the crew were German, including the Cook and second Cook and most of the seamen. There was also a young German cabin boy who could not have been more than sixteen years of age, and he was to be the cause of the biggest panic amongst the crew. The ship was tied up at a long jetty which stretched out from the nearest land, and the crew had to go along this jetty to get to the nearest pub. I was in my cabin about nine p.m, when I heard pandemonium on board the ship and the shrill blast of the ship's whistle piercing the air. Most of the crew and officers were ashore so it took some time to find out what was happening. Apparently one of the German sailors rushed aboard to say the cabin boy had fallen over the jetty at low tide and his head was stuck in the mud. He needed immediate help. Some of the crew were trying to pull him free and eventually succeeded and he was rushed to hospital, where he later recovered without any noticeable ill effects. I noticed the crew staggering back to the ship, much the worse for wear and there was a lot of agro in the air. There were scuffles, and Germans shouting everywhere. Some crewmembers were running and appeared scared. Then I saw Terry, our ex Irish Guard staggering, and mouthing choice language about 'Germans, Nazis, and snakes in the grass.' He wanted to fight any Germans, one at a time or all together. He was really fired up. At the gangway the German second Cook was trying to usher his countrymen to their cabins. If any put up resistance he delivered a beautiful right hook to their jaw and loaded them to their bunks. That fellow stopped a bloody massacre that night. Terry arrived back on board looking for someone to fight with and I went up to him and calmed him down. Sometime later he went to the loo and was found there snoring and fast asleep the next morning.

To find the cause for all the fighting we have to go back to the night before. Apparently, the German crew were in a pub having a few quiet drinks and singing songs, when Terry and some of the British crewmen arrived. A few drinks later a few verbal exchanges took place and all hell broke loose. Terry was up on his feet challenging any German to stand up and fight, man to man, or all were welcome. A few apparently took up the offer so the Brits and Germans had a free for all in the pub. At this point the young cabin boy got scared and ran towards the ship with some mates of his. He tripped and fell over side of jetty, which did not have any lighting. This is where this story began and fortunately nobody was killed or seriously injured.

Separate from this episode but on the same night the Scottish crew member who was an ex Black Watch soldier did not have the same results. He got seriously drunk and fell asleep at the side of the road that night. A couple of his buddies found him on their way back to the ship and tried to awaken him. He had been shell-shocked during the war and still suffered from this problem. When awoken on the side of the road he immediately flipped and kicked out breaking one of the crew member's legs. He later said that the fact Germans were aboard the ship put all kinds of thoughts into his mind and he believed he was under attack from Germans again, and he was reliving his wartime years. Poor fellow had to be sent home and the other guy was sent into hospital and then repatriated home. Some night! This ship and trip was becoming a nightmare and the worst seagoing experience I've had to date.

When all had calmed down the ship left Brest and headed for the English Channel to await orders. As usual, I was performing my radio duties when I got a call from Landsend radio saying there was a message for the ship. I took down the message and was shocked with the contents, so I asked for a repeat. It was still the same and I looked at it digesting the contents for the Captain, wondering what he was going to say or how he would feel. If it was for me I would be devastated. Anyway, I took the message to the

Captain and handed it over, keeping my eyes averted whilst he read it. I then asked if there was any reply. He smiled and said "No."

I said I was sorry. He asked me why and I said; "because they asked you to 'Proceed Falmouth and resign' I said. "Surely Captain, you're not happy being asked to resign."

He laughed; "The Company are saving money! Instead of paying for the two words re and sign, they have made one word out of two and it looks like 'resign.' I felt relieved and bit silly.

My voyage finished in Falmouth on the January 23rd, 1952 and I was glad it was over, even though I had made good friends onboard. The time I spent on this ship was over twelve months and that was enough. I now was due some extended leave and went home within two days.

While at home I had a great welcome and soon I was back into the swing of things in Cork. I longed for friends who were abroad to return home. At this time of the year it could be boring with no events worth mentioning, especially when I had to keep myself amused. One memory that does stand out is of daddy and me in Fota Estate on a lovely sunny day. He was enthralled with tales of my trip, which I edited for his consumption. I have a photo taken with him on that day and I treasure it. Soon after this I got my next telegram and returned to Southampton, on March 10th, 1952 where I was put on standby for my next ship.

Chapter nineteen

My next voyage took place on another T2 type tanker called the m.v. Esso Birmingham. It was built in August 1944 by Sun Ship Building and Dry Dock in Co. Chester, Pennsylvania and was originally named m.v. Maudville. Her official number was 181559 and she was a gross tonnage of 6324. She was later scrapped in Yugoslavia.

Between voyages there were two different Captains who shared command duties, namely; Captain W. Pittman, who commanded between March 14th to June 24th, 1952 and Captain R. Davies who commanded between June 25th and August 21st 1952.

The radio equipment on board this ship was installed by Marconi Marine Radio. This voyage was to be to the Persian Gulf, India, Southampton and Venezuela and back to Germany. When I signed on in Southampton, on March 14th 1952 I was told that the ship was due to sail for the Persian Gulf to load a cargo of crude oil for refining at Southampton. At this time I was not aware of the later trips to be undertaken on this ship and was a bit cheesed off at the thought of going back to the Persian Gulf. Uppermost in my mind was the memory of my last trip there and how lousy the weather was in terms of temperature and humidity. Anyway, there was nothing to do but concentrate on the good points - whatever they might turn out to be. Maybe the sight of glowing phosphorous in the water might brighten up my spirits. With these thoughts in mind I looked around the ship and was pleasantly surprised to realise that I was a board a tanker which was clean, and I later found out that the food was good by ship's standards. After meeting some of the crew one of the first things I learned was the interpretation the crew had for ESSO which was emblazoned on the funnel. They reckon it stood for 'Eat, Sleep, Shit, and Overtime.'

Captain Pittman was alright and we got on fine. It was a similar story for Captain Davies. As on all ships I reported directly to the Captain who in this case was Captain Pittman and I had no responsibilities for interfacing with anybody else.

When all the necessary preparations were ready the Pilot came aboard and, with the assistance of two tugs, the Esso Birmingham left Southampton, emerged into the English Channel where the pilot disembarked, and we headed for the Persian Gulf. Our journey first took us through the Bay of Biscay where we were confronted with a large swell and some rough seas but nothing like I had experienced on the Winkleigh thankfully. The rain came down heavily and there was no chance to go out on deck. I stayed in the Radio Room and when not on watch wrote letters or tuned in to short wave to listen to the BBC. There was plenty of time to stay out on the bridge and natter to the officer on watch or chat to the seaman at the wheel. Gibraltar soon appeared and we were through into the Mediterranean. As we cruised along the blue sea under beautiful sunny blue skies I marvelled at the sights of sea birds and marine life which abounded close by. This was a busy shipping area and there were many tankers, tramp steamers and passenger ships to be seen. There were many radio stations broadcasting sea traffic, and Alexandria Radio with its sing-song tone was quite audible and I contacted them to advise them of our destination and where we came from. It was not long before Port Said came into view and we waited for the Pilot who would guide us through the Suez Canal.

The shipping agent came aboard with our mail and I was pleased to get two letters; one from home and one from Molly. I was soon up to speed with happenings at home and got all the gossip. Molly's letter was personal and I read it over and over. How I wished she was with me to enjoy the trip. I had already written two letters and asked the agent to post them for me which he graciously did. I now started to write another letter to Molly and tell her how much I missed her and her smile. My mind was full of the lovely times we had and I felt from her letter that she too was feeling a bit lonesome. In the midst of this reminiscing I had not noticed that the searchlights were fitted and we proceeded with the rest of the convoy towards Suez and the Red Sea. As usual the British Tommies lined the canal bank and were ranting and cheering at the ships going by and we got loud cheers because the ship was clean. I saw the same sights that I had seen before and I was getting a bit bored. I was also suffering from a toothache and hoped for change in my luck. At Suez we again let the Pilot go and we sailed into the Red Sea. Into the dreaded Persian Gulf we sailed and on to Kuwait where we loaded Crude oil. This time the weather did not seem as bad as I expected and I found it bearable but the smells still lingered. I had not noticed on my previous voyage but there was a distinct smell of oil in the air.

    I went ashore to a dentist because of the pain I was suffering from an exposed nerve on one of my teeth. Lo and behold, the dentist hailed from Waterford! What I remember most about the visit was the treatment I received to alleviate the toothache. The dentist applied Silver Nitrate to the tooth, and told me it would not taste too nice but that the bad taste was temporary. What he did not tell me was that I should not smoke for a day after the treatment. My first inclination after being to the dentist was to light a cigarette and I took a deep drag. My stomach churned and I gagged. It was the worst feeling I had ever had up to that moment in my life. The reaction between the cigarette smoke and the Silver Nitrate was overpowering, and revolting. I could not enjoy a smoke for a whole day after the treatment. Somebody told me that this chemical was used as an aid to help those who wanted to give up smoking. A small drop touched to the end of a cigarette gives the same effect. I can well imagine you would sooner kick the habit than put up with the effects. Anyway it cured my toothache.

    From Kuwait, we called to Ras Tanura in the Gulf, and continued to load our oil before heading for a port close to Thane in India, where we tied up to a jetty off shore. There was no chance of going ashore as we were not there long enough, and there was no transport on the jetty so there was nothing exciting to report. All I can remember is that a number of locals were selling brass ornaments. There was a great demand for brass to make these ornaments so all means were used to obtain this metal. They were adept at using the bare soles of their feet to unscrew and remove the heavy brass screws which covered the sounding holes in the tanks that held bunkers. These are access points to dip and measure fuel levels. The brass that they absconded with would then be melted down and ornaments would be manufactured which they would then attempt to sell back to the very people they had stolen it from in the first place!

When we were finished we headed home, fully loaded, and deep in the water. In the bad weather which hit us on the way back the decks were awash and it looked like the ship forward of the bridge accommodation, was submerged. We had come back down the Gulf and into the Red Sea before re-entering the Suez Canal and after the usual fitting of lights and collecting our pilot we came out at Port Said and into the Mediterranean. Now we were back to the beautiful blue skies and blue sea which was one of my pet loves. Gibraltar rose up in front of us and we steamed through and headed for the Bay of Biscay. Again, luck was with us and the weather, though dull and misty, was quite calm. In no time we eventually arrived back in Southampton, after an uneventful trip, where I signed off again on May 8th, 1952. I re-signed on the next day for the continuation of the voyage.

This time we sailed across the Atlantic to Caracas in Maracaibo Bay, Venezuela. This Bay was hot and humid, and very uncomfortable. It was interesting to see the Pilot navigating through the vast arrays of drilling and pumping rigs in the bay. The things looked like huge iron birds, feeding by going down and up with their beaks. There was no stopping the activity of these 'Iron Beasts.' We pulled alongside the wharf and tied up. The hosepipes were connected in no time and oil gushed into the tanks. The Captain always wanted a quick turn around.

When we got a chance a few of us headed off ashore, we did not have much time to browse as the oil was gushing into the ship at an alarming rate. The closest area was well lit with neon signs and there was plenty of action. The chatter and noise was deafening. I walked around for a while and bought some souvenirs, and I decided to buy some coca colas with the crew members. We sat and drank the cokes for a while, conscious of a number of locals who were carefully watching us. We began to feel uneasy and decided it was best to head back to the ship whilst there was still light. At this time the vigilantes were out in force, but we only found this out later. We could easily have been thrown in prison for any disturbance or any pretext. We casually started to walk back to the ship, watching the fellows who had left the bar behind us. They did not look too pleasant, and when we got within sight of the gates we ran for the docks as fast as we could. No more shore leave, but in any case we did not have a chance, as we left soon the next day and headed back to Europe.

On the way back I saw a number of dolphins which raced ahead and criss-crossed our bows. Further out in the Atlantic, at sunset, I watched varying shades of colours as the sun sank into the western horizon and at night I watched the stars shining brightly and some shooting stars light the sky. There were particularly spectacular views of the sky as lightening flashed through the dark night. This lasted for about an hour and became a memory which was refreshed on later voyages.

Amidst all this the ship cruised on towards the English Channel and I kept taking weather reports and navigation warnings as well as keeping watch on the distress frequency. There was one distress call from a ship which was too far away for us to render assistance. Due to the weak signals and the fact that other ships were in communication with the stricken vessel we did not need to assist. I had to maintain my own watch and did not have the authority to send signals in case of disturbing the rescue, and when the 'all clear' was signalled to advise that the distress was over we were all relieved. The local coast stations returned to working normally and I went back to listening to Landsend Radio traffic lists. I heard my ship's call sign and received a message for the Captain. This was one of a number of messages I sent and received during the voyage and like all the received messages they were for the Captain and related to ship's business. Soon, we were through the Channel and at our destination where we picked up our

Pilot who guided us to Imjuiden and Rotterdam, in Holland for discharging our crude oil. We tied up at the wharf where hose pipes were connected and oil began to flow ashore into the tanks. The whole operation was quite noisy and I was glad to be able to get ashore for a brief respite. Rotterdam looked wonderful and I wished for more time to rummage around and see the city.

As usual, I went shopping and not much else. I did not even have the time to walk around or see the sights. I'd have another chance to visit Rotterdam later, on another ship, but was not aware of this at the time. The oil was discharged by the efficient Dutch dockers and the Pilot boarded before the ship left Imjuiden and disembarked the Pilot and headed back to sea for another trip across the Atlantic, for another cargo of oil for Southampton.

Like the last part of the voyage there was nothing different to before except that I did not bother going ashore and spent most of my time reading and writing letters as well as re-reading letters from home. Of course I had some nice Spanish music to listen to as well and spent time chatting to the crew. Again, we set sail for Southampton where we arrived without mishap and went through the discharging process. I signed off here on June 24th 1952 but had to re-sign on again the next day to complete the contract. Our orders arrived with our mail from home and we headed back to sea. We arrived back in Rotterdam with another cargo of Crude oil. Our next port of call was Hamburg in Germany. Having dropped the Pilot we set course up the North Sea towards Hamburg.

We picked up the German Pilot and proceeded up the Elbe to Hamburg. I went ashore to St.Pauli. Out of curiosity, I went into same bar where I had my escapade on the Orford. Whilst I was drinking my coke and looking out the window, I saw an American jeep of with military police as it pulled up outside the Bar where I had been duped. My curiosity got the better of me, and I went outside to see what was happening. After a while I saw the Police coming out, and they took a number of girls with them. I could not believe it when I saw the girl who'd tried to cheat me amongst these girls. I waved furiously, with delight, as she passed by in the jeep with the Police. I'm not sure if she saw me but I saw her. Boy, did I feel good seeing her get her comeuppance and I returned to the ship a happy man. I would not have missed it for anything. I was still enjoying the episode when on return to the ship I was informed that I would be flying home within a few days.

    The ship was due for a spell in dry-dock for overhaul, so we set off down the Elbe, where the Pilot disembarked, and we set off for Bremerhaven, Germany. On arrival we signed off, for the last time, in Bremerhaven, Germany on August 21st, 1952 and I flew back to the UK where I reported to the Southampton office. I got another warrant and I got home for a short spot of leave.

    I had bought my first car, a Hillman Minx, some years earlier and decided I would use it for trips to Cork. I had left the car at home while I was at sea and daddy was using it. The next thing I heard was that my car had been raffled in a draw which daddy had agreed with Perks' Amusements that were in full swing in the Bath's Quay. The car was on a stand specially erected for the draw. Daddy had his quota of tickets but did not win. How much he got I don't know. Anyway, when I came home on leave I did not have a car or the sale value of it. I soon forgot about it and told him he could keep the proceeds. I still don't know to this day how much he got or who won it. I now had to depend on the train to get to Cork and this governed my time available as I had to get the last train at 11.15 P.M. I continued to meet Molly and we carried on as usual, comfortable in each other's company and fed up when we had to part. However, there were still no thoughts of marriage by either one of us at this time. As usual, my family wanted to hear stories of my life at sea and the places I had been. Daddy was like a schoolboy checking anything new I might have from abroad. Before long before a telegram arrived, asking me to report back to Hull to await my next assignment. I was in for a lovely surprise with the voyages I was about to partake in but had to await arrival in Hull to find out.

Chapter twenty

When I arrived in the Hull office I was sent to West Hartlepool to join my next ship, which was my eighth ship and was named s.s Sedgepool. The Master was Captain J. E. Roclar. This was a Liberty built ship, built in 1944 by the War Shipping Administration, for South-Eastern S.B Corp. in Savannah, Georgia in the United States. Her Gross Tonnage was 7278. I signed on in West Hartlepool, England on September 15th, 1952.

My seventh deep sea voyage was to the USA, Panama, Honolulu, Japan, Canada, Mexico, Peru, and Italy. This was an eventful journey from beginning to end. From the day we came down the North Sea and into the English Channel before emerging into the Atlantic Ocean we did not realise the problems that lay ahead. The sea traffic was quite heavy. Soon after we passed the southern Irish coast we ran into quite heavy seas. The Chief Mate was on deck a lot when he was not on watch. The Chief Engineer was usually with him, and then I began to realise what they were doing. Drawings were being scanned and discussed, and wooden planks were neatly stacked on deck. Gradually, during the trip the outline, and finally the finished product, of what they were working on appeared in all its glory. A beautiful twelve foot clinker boat had been built on board the ship. The trip was quite long and the Mate had ample spells of time between watch keeping and poor weather conditions, to get his pet hobby finished. The Mate was justly proud of his work.

     The weather improved for a while, but the Captain requested that I kept taking all the weather reports I could find, as we were being warned of a severe storm forming ahead of us. The Captain sent regular coded weather reports to the metrological office, outlining prevailing conditions where our ship was located. There were other ships at various points at sea, and these performed the same function. As we progressed, the Captain read one of the messages I had received, looked perplexed, and asked me to get a verification of the position of the centre of the storm, as he did not believe it to be correct. I immediately contacted the coast station, and the message was repeated to me. It was identical to the original message.

The Captain immediately asked me to standby for a priority message. Without delay, he took all readings of atmospheric pressure, wind direction and force, longitude and latitude, and various other required data, which he then coded into five figure groups. Looking out the bridge window, the sea was dead calm and not a puff of breeze was visible. The sun shone brightly and I could not understand the anxiety of the Captain. When I got the message from the Captain, I immediately sent this off to the metrological office. A short time later, Landsend Radio sent a TTT message, which indicated a navigational warning to all ships. This message was from the Metrological office, warning all ships that the centre of a severe storm was now located at the precise longitude and latitude that our Captain had given. We were smack in the middle of this storm and would soon find out how strong it was. We were in the eye!

The crew started battening down the hatches and all loose deck equipment, including the awnings. The mate moved his wood and tools from the deck, and we awaited the oncoming storm. As we altered course to avoid the worst of the storm I felt the breeze stiffen, and black clouds began to fill the sky. Gradually, the sea began to get very angry, with white-capped waves getting bigger by the hour. Heavy rain descended and lashed into the ship. The wind was increasing in intensity, and anything not tied down was thrown around the cabins. Lightning flashes were regular and lit up the sky. At times I felt the claps of thunder and the lightening occurred at the same time. It was frightening to say the least. The Captain slowed the engines and tried to ride out the storm. It was miserable and very dangerous because we were in a severe force ten storm. I wondered how you could go from a sea as calm as glass with no wind into such a ferocious storm, in such a short space of time. The waves must have reached forty feet. I could not hold a steady position in the Radio room. My chair kept sliding and twisting with wave after wave, and when the ship rose on the crest of a wave, I waited for the pounding as it descended into the trough. It reminded me of my trip on the Winkleigh through the Bay of Biscay. At night it was chaotic trying to stay secure in the bunk and I got precious little sleep. Fortunately, I was not one for getting sea sick, and eventually the ship rode out the storm. There were no casualties, and the ship survived well.

The relief, as we eventually nosed our way out of the storm, was glorious. Gradually, we steamed our way into blue skies and a gentle rolling sea. With the clear weather back, the Mate got his boat out again...

About a week later the second steward got ill. He had a very high temperature and a very bad rash all over his body. The Captain and Mate were baffled and looked up the medical book with no success. Next stop was to my office, where the Captain asked me to contact any ship with a doctor on board. I immediately sent out a 'Medico' request to all ships and got answers from the Cunard liner Queen Mary and another passenger liner. I started to work the Queen, which was closer. We exchanged Morse messages, as our ships did not have Radio Telephone facilities. The Captain gave details of the symptoms, temperature, pulse rate and various other bits of information requested by the doctor. The doctor then diagnosed possible Nettle rash. Whilst I was in communication with the Queen Mary, another call came into me from Rome through the local Radio station (IAR) requesting that I communicate directly with their international marine medical facility. I thanked the doctor and Radio Officer of the Queen Mary and carried on a working frequency with Rome. Their advice was to use Penicillin injections, which we had in stock in the medical locker. The Chief Mate delivered the necessary to the patient who brightened up after a short while.

Now we had another problem. While the patient was comfortable, the Chief Mate began to react to the Penicillin. He was allergic to this medication and may have got some on his hands. Medical hygiene was not the really practiced in those days. His face swelled up and his eyes closed. He was having difficulty swallowing. He looked blotchy and red faced. I was again back on the Radio. I made direct contact with Rome again and gave information on the symptoms. Their response was immediate, clearly identifying the Penicillin as the problem. Thankfully, with their intervention, both patients recovered with no ill effects. I was delighted to have been able to justify my job. I could have gone for years without doing anything outside routine duties.

The first day after we left port I encountered a technical problem with the transmitter which needed attention. There was a motor alternator which was used to start-up and supply power to the transmitter. It was essential that this equipment worked well, otherwise there was a possibility that when the need arose this piece of equipment might not work and this could be catastrophic. From day one, I noticed that even though it was working, there was no guarantee that it would continue to work successfully. I suspected it had to do with the copper rings, called slip rings, which were part of the alternator and that these might be worn in places. There were fluctuations in meter readings and instability in the sound of the motor. I spoke to the Chief Engineer and explained the symptoms to him and he took the alternator to the Engine room and did a fantastic job on the rings. The rings were in fact egg shaped so he put them on the lathe and ground them into completely circular rings and it was a pleasure to have a transmitter that worked 100% again.

Up to now, we were heading west awaiting orders, and when they came I got the biggest shock and thrill I had ever felt at sea. We were to proceed to a place called Kamaishi in Japan. Charts were taken out for the Pacific Ocean and we all looked to see where Kamaishi was located. It is a small port in the northern end of the main Island of Honshu. It was apparently a coal town. This was going to be a long voyage across the Atlantic, through the Panama Canal, and through the Pacific Ocean. Is it any wonder these ships are called 'Tramps' when they depended on trips like this to pick up a cargo? It was a case of put to sea and await orders. Something like a lottery and nobody knew where the ship would end up.

With a couple of new first experiences looming for me we headed for the Panama Canal. Since I had not been here before I was looking forward to this venture. It was glorious going south into the lovely sunshine, watching the ever present slinky dolphins criss-crossing the bow of the ship. Most of the time the sea was calm, and as we passed close by land we could see the pelicans, and other birds, flying around and squawking. We passed close to other ships, including passenger ships that were trading between the Caribbean Islands and also heading for the Panama Canal. We arrived at the Canal Atlantic port of Colon, picked up our Pilot and entered the Canal as we awaited our turn to pass through the first of three locks, which were operated by Americans. The Canal was so different to the Suez Canal. In Suez there was no sea level difference between the Mediterranean and the Red Sea, so there were no locks to raise or lower the ships. Here, there was a big difference in levels between the Oceans. Three locks were in operation. They were in 2 step flights at Miraflores and 1 at Pedro Miguel.

We entered the locks, which were then closed behind our ship, and the water level was adjusted rapidly to flood the lock, or empty it, depending on whether you were going towards the Pacific or towards the Atlantic. Looking over the ship's sides as this was happening there were various types of fish sucked into the locks. There was a peculiar sensation similar to the Boal, in the Seine, as the ship rose or fell rapidly with the water level changes which occurred. When the correct level was achieved the lock's gates were opened, and the ship moved towards the next lock. She was then towed by what were called 'Mules.' These were mechanical type tractors, which moved on serrated rail tracks at each side of the lock. They climbed up to the next lock level, towing the ship into the lock. It was necessary to use three locks due to the water height difference between the Atlantic and Pacific. One lock would not be able to cope with this difference, so it was done in increments using the three locks. It was a marvellous piece of Engineering.

The trip through the Canal was lovely, and there were men at the side of the locks using high-pressure hoses to cut away more rocks and silt to widen the canal. I did this trip many more times and was always in awe of the feat performed opening this canal. It is certainly, to me, one of the Great Wonders of the World.

We came out at Balboa and headed across the blue Pacific Ocean. This was my first trip to this side of the world and all I could see were blue skies, sunshine and lovely blue seas. I thought 'this is the life and I'm getting paid and fed for doing this. Yes, please!' I did a fair bit of sunbathing while we moved towards Japan but there was still one more port to call to in order that we got bunkers… Honolulu. This conjured up lots of fantasy thoughts in my mind like beautiful maidens in skimpy grass skirts dancing and putting garlands of flowers around my neck. Unfortunately, we would not be there long, as we only needed to take on fuel and some fresh food. I was not disappointed when I saw the island of Hawaii appear on the horizon, and I kept looking through binoculars to see if I could see any fair maidens in grass skirts. Alas, no!

We tied up and the ship's agent came aboard, delivered and took our mail, and asked me if I wanted any spare equipment for the radio room. I didn't, but I could not refuse an excuse like this to get ashore and said I would take a few spare valves if Marconi had an office there. I was taken ashore and got my spares. The Marconi representative asked me if I had been there before and when I said no he offered to give me a whistle stop tour in his car. It was a very pleasant surprise. I saw beautiful grass parks still suffering the after effects of the Japanese attack shown on the film I had watched as a boy; 'Tora, Tora, Tora.' The craters were still visible and so were the large oil drums scattered around any flat area where planes might have landed. He told me that during the attack by Japan, there was the fear that the Japanese would try and land planes to invade the islands. Trenches were dug in the flat areas and any large objects placed strategically to hinder planes from landing. They were still there in some places while we were there. He also took me to the market where all kinds of exotic fruit and flowers were being sold. Of course, I had to see the after effects of the horrible bombing of the American Fleet. I saw some of the wreckage but I did not have time to take it all in as we were about to leave and head for Japan.

After an uneventful journey, as we closed in on Japanese land excitement began to rise. What would it be like? How would we be received? After all it was not that long since the war finished and we assumed that resentment must still affect the people's thoughts. Soon we would find out. On arrival in Kamaishi, we entered a very small port and tied up at the dock where there were a number of coal grabs. The town was dull and dingy looking. I remember one main street and the people looked depressed and certainly I got the feeling that they resented any Westerners. The contrast between the men and women was immediately obvious. The men were sullen and either lounged around, or formed groups who viewed us suspiciously. I did feel uncomfortable but not intimidated by this. The women were subdued and were the ones who had pickaxes and shovels and were digging the road. From what I saw, they did all the hard labour while the men relaxed and lounged around. The women usually kept their heads down, and would not glance up when we passed. On the other hand, I saw the other side of the women who wore Geisha outfits, and who shuffled along beautifully painted but wore no smiles.

On our first evening ashore I was with a couple of the lads and we went into a Japanese bar. There appeared to be quite a bit of activity, as there is in most ports around the world. We opted for a separate room and as we entered we were requested to take our shoes off and leave them outside the fragile door. Two of the three of us had holes in our socks and felt mortified. I was never again caught with holes in my socks, at least in Japan anyway. When we entered this small room, the partitions appeared to be made of very light, semi-transparent material. The top of the table was very close to the floor. We had to sit cross-legged around this whilst a Geisha girl waited behind each of us. The young girl who attended me was named Yoka. As usual, I wanted a Coca-cola, and she would scuttle off to get me another bottle when I needed it. She was a pretty, petite little girl, about five feet tall, and beautifully made up and robed. She could not speak English so I could not chat to her, but one of the girls did have a little English and I learned a few phrases of Japanese.

While we were in this room I wanted to go to the loo, so I asked where it was. The English speaking girl said something to Yoka who got up, stood back, and beckoned me to follow. She led me to a small toilet, where there were no sit down facilities but did have the usual gent's wall to go against. I went in and Yoko went elsewhere. I did my business and was closing my fly when I turned around there was Yoka right behind me, holding a basin of water and a towel. I thought, 'What is she going to do now. Is it the custom here that she should wash me after I use the toilet?' Anyway, she pushed the bowl towards me and I got the message that it was for me to wash my hands in, and for no more exotic purposes than that. I couldn't make out if I was relieved or disappointed at this realisation.

We went back to the room and rejoined the lads. Things appeared to be livening up, and the two Geishas were smiling and enjoying themselves. Yoka never left my side. She would smile every time I looked her way - which was quite often - and I felt that the white makeup would crack on her face when she smiled. Once, when she brought me a Coke, I kissed her cheek and she blushed, and then smiled. After that I could not move without her following me everywhere. One evening I was returning to the ship on the main street and noticed Yoka was right behind me. I did not expect her to follow me back and was shocked when some of the layabouts on the street started to yell at her and scorn her. I felt embarrassed for her and asked her to go back. She did not understand and followed me to the ship, and only after I had boarded did she return back to the eating-house.

Three of us went into every bar in the town; these could be easily identified because bunting hung halfway down to a half door. The door had to be pushed open, and as we walked through the bunting - presto - we were in the bar. Some of the crew were getting up all kinds of capers in these bars, so we did not hang around them too long. However, one evening we were walking along the main drag, and I spotted a Geisha girl going through one of these doors where we had not been. I ran across the road with the lads in tow, and straight through the door and bunting. I only just stopped in time, as my foot was only about three feet from the edge of a bath, full of nude Japanese women. We had accidentally stumbled into a public bathing area for women. The women were everywhere. Some were in the pool washing themselves, more were outside the pool drying themselves, whilst more were undressing or dressing. As we came to an embarrassing stop, all the women turned to face us and bowed, and then carried on what they were doing. It did not seem to create any problem for them. It did for us! Whilst we looked, and tried to gather our wits, I noticed that there was a wooden partition between this room and the next one, which was the men's bathing pool. This partition had a gap of about one foot above water level, so that it was possible for the men and women to see each other. We got the hell out of there after what felt like an hour, but in fact were probably only minutes. It was some experience. Still, just one of many unusual ones I had in Japan.

After a number of days we finished loading our cargo of coal and headed back across the Pacific. Our next sequence was to the West Coast of the States, where we berthed first in Seattle, Washington and then in Portland, Oregon. I remember trying to fish for salmon in both these ports. I used a rod and line with a spoon but to no avail. Other fishermen were catching lots with drift nets, which I was told were illegal.

I spent a fair bit of time looking around both these cities and was mainly on my own as I had the time off. On the coastline there was snow on the hills and this made it very similar to the Norwegian coast. The whole scene was beautiful and everything moved at a gentle pace. Having discharged our cargo we loaded grain and carried on to Vancouver, Canada, which was right across from Seattle, where we loaded more grain.

I was enjoying seeing so many new places. In Vancouver I went ashore as much as I could but due to the speed of loading the grain we were ready to pick up our Pilot and put to sea after a short stay. I was a bit lonesome leaving and despite the short stay felt that I could have happily lived here. The people were nice and I saw Mounties (Mounted Police) for the first time. It was very similar to the cities in the States but yet there was something different….

From Vancouver we headed still further north to Prince Rupert Island. When I heard the name 'Rupert' my memory immediately jumped back to my childhood, when my mother used to read 'Rupert the Bear' stories to me and my sister, which we loved. Prince Rupert Island is a small island located 74.54.09N 130.20W between Queen Charlotte Islands and the West Coast of mainland, Canada. As we threaded our way through narrow inlets it was awe inspiring to see the snow-capped mountains on the starboard side, close by, and more snow on the port side. It was beautiful, clean, cold and very fresh.

Prince Rupert was a busy little place and was patrolled by the Mounted police. On my first day, as I leisurely walked along the street, I started to cross the road without thinking, or looking right and left.

"Hey, you come here," shouted a voice, which belonged to a pimply-faced Mountie.

I turned, and he asked me who I was and what the hell I thought I was doing crossing against the traffic lights. I looked around but did not see any lights and told him so.

"Do you ever look above you?" he asked.

There strung across the road, high above my head, were the lights. I had never seen this type of light set up before. I apologised and he told me to watch it in future, so I took his advice in Prince Rupert.

Later on, I strolled across a square off the main street and heard female voices calling. I looked up and saw a number of women shouting and making rude suggestions with their hands, as they peered through barred windows. Later, I learned that this was a holding prison for females. They sure had great fun when men passed by. We completed loading the rest of the grain in a day or two and headed south.

Again another bit of bad luck for this ship, and it was not to be the last. As we travelled down the West Coast of Canada the weather began to deteriorate and we had to batten down hatches and secure all loose fittings, as the forecast was not good. We hit the storm head on and battered away against it. Everything looked OK as the weather improved. Then the Bosun noticed that there was something wrong in the after hold. Water had got into the hold. The wheat stored there had begun to swell and combustion was a distinct possibility. The hold was opened to reveal that this indeed was the case. There was an enormous swelling taking place in the hold, and heat was being generated so there was a distinct possibility that either it would explode, or damage the sides of the ship if the problem was not solved. There was only one thing to do, and that was to start discharging the wheat and throwing it overboard. All hands were called to deck, and the slow process of removing the grain began. We seemed to be making little progress, and I was asked by the Captain to send messages to our shipping agents in Oakland, San Francisco for permission to enter port and discharge the cargo. We got the affirmative, and when we got tied up all the usual Customs, Doctor, and Immigration, as well as Fire Appliances, were all waiting. When the shore firemen checked the situation we were ordered to immediately leave port for their safety's sake, as they confirmed that there was great danger of an explosion on board and the Americans were not taking any chances.

The Captain had no alternative but to put to sea again and hope for the best. We were all sweating about our predicament, and we were doing our best to heave the offending cargo overboard. As we sailed south we had refusals from San Diego and San Pedro, California to let us enter their ports, so we had to continue south for a few days until we got to Mexico. Here we got permission to enter Guaymas, Mexico in the Gulf of California, and they discharged our cargo. We all breathed great sighs of relief because we never knew who amongst us could have been killed or maimed had the combustion and explosion taken place. I never got a chance to go ashore in Guaymas and spent the short while in port fishing. When we left we went to Mazatlan further south in Mexico, where I did some great fishing from the side of the ship. It reminded me of fishing for mackerel in Cobh because the fish gave themselves up easily. The only problem was that amongst the lovely snappers, there were eels which curled and twisted, making one hell of a mess of the line. There were also fish that were known locally as 'Pigfish'. These were a kind of brown fish, with yellow stripes, and they looked like pigs. They were round, and could blow themselves up to twice their size and then fire water out at you. They were horrible, and I used to cut the line rather than try and unhook them.

We left Mazatlan and Mexico and sailed to Lima, Peru where we were to fill one hold with cargo of ore. Here, I was offered a job by a fellow Irishman, to act as supervisor in a mine. The pay he assured me would be very good and he would organise my clearance to jump ship. I turned it down for obvious reasons. I also met Irish priests who were in the town, which was nice. The people appeared happy but poor. They were dressed in multi-coloured clothes; this was the land of the Incas.

This time we had to go through the Panama Canal from Balboa on the Pacific to the Atlantic. We left Colon and we arrived in the Atlantic side of the world. We headed north to New York. The weather was fine as we passed through the Gulf of Mexico and up past Miami, Florida. When we hit Cape Hatteras we again encountered heavy weather. We had this bad weather for a few days and eventually came to the Ambrose Channel in New York, and the pilot guided us to our berth in the centre of New York. It was March 1953.

The entry into New York was terrific. The whole outline of Manhattan Island and the Statue of Liberty were sights that remain with me to this day, and I thought of the poor Irish Immigrants who came here in Tall ships, and landed in Ellis

Island. A lot of them would have come via my own hometown of Cobh. I was very disappointed that we were leaving on the 13th March, and so would miss the St. Patrick's Day parade in New York. The ship's cargo had been quickly discharged but I had time to browse around as we went to dry-dock for minor repairs. The shops amazed me and I ended up buying some clothes and shoes. I wondered what they'd think at home when I wore the midnight blue suit with the Orange and yellow tie! My feelings were that Molly would be horrified and I dared not mention it in my letter which I was about to post.

I would have liked more time here but I need not have worried about that. I know the 13th is supposed to be unlucky for some, but on that date it was very lucky for me, though unlucky for the ship's owners. On Friday the 13th March 1953 we left the wharf under the guidance of the pilot. I was in bed when we cast off. It was early in the morning and I was uneasy because I could hear ship's horn being sounded. I got up about six-thirty a.m. I looked out the portside porthole and saw thick pea soup fog and nothing else beyond the side of the ship. We did not have radar, and I wondered why the pilot had our ship under steam. I got dressed quickly, feeling very uneasy, and went into the radio room. The radio room was on the port side of the bridge, just behind the wing of the bridge. I switched on the radio, tested the equipment, and tried to relax, but the constant sounds of ship's whistles and hooters so frequently unnerved me. Some sounded very close and the ships could not be seen in this fog.

I looked out the porthole of the radio room and saw the outline of the 'American Farmer'- a large ship- pass within twenty feet of our ship. The ship's radar was in operation. There was constant hooting from us and other ships in the Channel as we headed towards the Statue of Liberty, all trying to notify each other of our presence to avoid collision. Then, all of a sudden, there were a series of sharp loud blasts from directly in front of us, and I heard the pilot shout "hard to starboard" at the same time blasting our horn in a series of quick blasts. My head was half-way out the port hole when I saw this huge towering bow of the Cunard liner 'Parthia' slide past us, metal to metal, at the same time as its anchor was ripping the port wing of the bridge away, which came towards my head as it began to collapse. The anchor then caught the lifeboats on the port side. It was around 7.30 am. There were sparks flying along the entire length of our ship. There was no room to put a playing a card between the sides of the two ships. The Parthia was huge compared to our little ship. It was a miracle that nobody was hurt, or that it was not a head on collision because surely we would have been holed. There was great credit due to the seaman on the wheel for keeping the ship steady when he saw the oncoming bow emerge from the fog, towering over our ship.

We immediately stopped engines and dropped anchor, whilst the liner kept going at a fair old speed, never slowing down. I started my transmitter and tuned my receiver; waiting for what I knew was going to be a message to the agents advising them of our accident. It was not long coming, and soon we had a tug alongside to tow us back to where we had just left, into Bethlehem Steel Dockyard in Brooklyn, New York.

That evening, whilst we were awaiting the ship's surveyor to come, I was standing on deck just looking around the place, when I saw this Merchant Navy Captain with his sidekick approaching our ship. They were the typical type of British officers that you would see in films - arrogant and full of their own importance.

They came up the gangway, making sure their uniforms were not soiled by touching the rails, and stepped on deck. I was speechless, and so were our Captain and Chief Mate, when the boarding Captain introduced himself as the Captain of the Parthia and his companion, the Chief Officer. It was his next words that threw us all.

"You must be the Captain of this heap of rust," he said addressing our Captain.

I thought our Captain would have a fit, or that the Chief Mate would clobber him. They all retired to the Captain's quarters for a 'meeting' and I daren't imagine what was said there!

1951; ss Orford. My 6th ship

1951; Father and son at Fota Island, Cork

1952 ss Sedgepool My 8th ship

1952; ss Sedgepool. More collision damage

BEYOND THE SEA

*1952; ss Sedgepool Collision damage in New York
(Both Images)*

We were soon into dock and repairs commenced after surveyors checked and assessed the damage to our ship. I had ample time to get really excited and prepare to walk up to see the Parade. It turned out we were to spend many weeks in New York. I went ashore for periods during the days as I eagerly waited for March 17th, St. Patrick's Day. This, I expected to be a momentous day in my life. How many at home would ever get the chance to be physically present in New York on this holiday? On the big day I got up early, shaved and decided I'd have something to eat on the way to 5th Avenue. Whilst I was ashore I had mentally mapped my route and was very excited to see green streets and green flags. On this morning, as I strolled towards 5th Avenue, I called into a smallish restaurant that was decked out in green. I sat down and a waitress came up to me and handed me a menu. She too, was dressed in green. I had never seen anything like this, and wondered what it would be like to be a ringside viewer of the actual parade. I ordered rashers, eggs, hash browns, potatoes and a cup of coffee. The waitress looked at me in a strange way.

"Sorry but I won't serve you that meal," she said with a smile.

I was flabbergasted. As she continued to look at me I started to get annoyed. Was she going to spoil my day I wondered?

She said "Try again."

I said, "It says on the menu that you serve the meal I ordered, so why can't I have it?"

She gave a sigh, looked me straight in the eye, and asked;

"Are you or are you not an Irishman?"

To which I replied, "Yes, and a genuine one at that, so can I have a breakfast please?"

"I'll tell you what I'll do for you," she said, "seeing as it's St. Patrick's Day, I'll bring you a true Irish breakfast and dinner, which we call Brunch."

She took off laughing. I had no idea that brunch was a combination of breakfast and lunch that Americans had invented. Sometime later she arrived back to my table, and I was gob smacked when she put a huge plate full of corned beef, cabbage, and potatoes in front of me.

"That's what all good Irishmen will be eating this day and it's on the house for you."

I could not get over the hospitality of this girl, and the way she enjoyed having me on. Neither can I forget a brunch such as this. It tasted lovely, and I certainly did not need to eat for the rest of the day. When I thanked her, she wished me well and told me to enjoy the day, and to call if I was ever in the vicinity. I never saw her again.

As I trundled up 42nd Street I could feel the excitement rising. The noise was getting louder as more and more people chatted, laughed and generally enjoyed themselves. Fortunately, it was a beautiful sunny day. Everywhere was awash with green. I think I was the most conventionally dressed that day. Even Japanese, Chinese and African Americans were all Irish for the day. The New York Police were everywhere, controlling the thousands that stretched all along 5th Avenue. I picked a spot near St. Patrick's Cathedral and took in the atmosphere and wondered if I would ever be back here again. In the distance I could hear the bands playing, and gradually they approached and my excitement grew. I started to fidget and look for a better position to view the marching groups. Marching up the centre of Fifth Avenue were various nationalities of people, and various bands - Pipe bands, Brass bands, Calypso bands and God only knows what else. Prominent were the New York Police, New York Firemen and various Army, Navy, Air Force, and Marine personnel. It was strange to see these military people walking out of step, and casually waving to the people on the sidewalks. Of course there were lots of groups singing and dancing as they passed by. I was really fascinated and enjoying all of this. All I could hear most of the time were the strains of old Irish songs and music. Then along came what I was told were the 'Black Irish.' These were Black Americans, who were Irish for the day and wore Green outfits. They were a sight to behold, and credit to them they really entered into the spirit of the festival. The parade continued for hours, with no gaps between groups that were usually eight people wide across the Avenue.

It was only after the parade finished, and I started to return to the ship, that I realised how tired I was from standing all day, and I now had to face the walk back to the ship amidst the throngs of people still celebrating. When I got back, late in the evening, I just sat down and thought of what a wonderful day it had been and how lucky I was to have been there. I suppose I have to thank the Cunard Line and their liner Parthia, not to mention the fog for all of this.

A few days later I got a letter from home telling me that a friend was married and living somewhere in the Long Island area. I was given a phone number and told to give her a ring. I contacted Sheila, who hailed from Cobh and was a sister of my friend Dr. Chris Walsh. I went to see her and we had a great chat, and then she asked me out to the house the following Sunday, as the local townspeople were holding their St. Patrick's Day Parade then. They had postponed it on the 17th so that they could march in the New York parade. I duly arrived there and at three p.m. that afternoon found myself leading the local St. Patrick's Day parade. I felt like a twit, out in front, followed by a statue of St. Patrick borne by four men. Here I was in navy blue gabardine and kid gloves, leading a crowd of fervent Holy Communion children, priests, police, and people dressed in Green. It was not something I would like to have happen again but on the day, as I was told, "You're a true Irishman living in Ireland - who better to lead us?" The parade was over after a short while, but memories of it burned in my mind for a long time after. Unfortunately, I never got a photo or kept the local press cuttings of it.

Most of the time I spent on my own, moving around New York, and found it to be a very lovely place full of vitality. The stores were huge, and very well stocked, and I had not seen anything like this except in the movies. Alas, all good things come to an end, and with repairs completed we eventually set off back down the Ambrose Channel, without any more accidents or mishaps.

Out into the Atlantic again, we headed south towards Miami, Florida, where we were bound for Corpus Christi in Texas. The weather was very hot around Miami. As we approached I had the binoculars out to see it all as we passed very close to the coast. Like in the movies, the beaches were white sand, crowded with all shades of humanity. The buildings stretched across the whole seafront and the cars were something else; open top limos with the most beautiful women driving around in them. They wore the skimpiest of bathing gear and most were beautifully suntanned. We were all mesmerised. Nowadays all can be seen on television or in the movies.

We carried on through the Gulf of Mexico and watched the dolphins surfacing and swimming in front of the ship. The turtles were doing the usual. No wonder the poor female had so many eggs to lay when the time came and she swam back exhausted to the sea. Gulls and pelicans were everywhere, and many fishing boats criss-crossed in front of us. There were a lot of ships and some beautiful yachts cruising around. This was what life was all about.

We sailed on for Corpus Christi in Texas. Here, our visit was quite short. We went to collect a cargo of grain, which was done in extra quick time. Instead of the old crane and grab, we were loaded with a large pipe from a silo, and the grain flowed like water. I spent most of my time on the beautiful beach watching the speed boats and yachts. When loaded, we were guided out by the pilot and left Corpus Christi and headed into the Gulf of Mexico, through the Straits of Florida, past Cuba, and set off across the Atlantic. This time we were headed for Genoa, Italy.

This was a long trip, as we still had to go through the Straits of Gibraltar and into the Mediterranean Sea. There was nothing exceptional about this trip across the Atlantic other than the beauty of the other ships lit up at night, passing us in the other direction, heading for the Caribbean or USA. We had been in the hurricane season but any warnings fortunately were further behind us so we avoided them. The ships passing going the other way may not have been so lucky.

On arrival in Genoa everything was ready for immediate discharging. The dockers were not long hooking up, and they started discharging the grain almost immediately. While this was going on I went ashore and ate some lovely Italian food. This was my first taste of this cuisine but I had to have coffee as I still had not started drinking alcohol. Ashore, near the docks, it seemed like any other port, with dockers, guards and general noise. I was getting tired of this ship and we were now nearly eleven months away from home. A long spell like this was eventually catching up with me and, thankfully, we were ready to leave in a few days and sail for home waters.

  We steamed out of the Mediterranean, into the Atlantic, and towards the Bay of Biscay. We were to have another baptism as we entered the Bay. The weather was foul and we got hammered for two days. 'Behind every cloud is a silver lining.' Joy was sublime as the storm abated and only a few days were left before I could get off this ship and get home to Cobh. Local radio signals Landsend (GLD), Niton (GNI), Northforeland (GNF), Cullercoats (GCC) and Humber (GKZ) were all loud and clear, as we sailed up the English Channel, round Northforeland, and up the North Sea, where we docked in North Shields. It was over eleven months since we left West Hartlepool, not far from this port. I was glad to see the Chief Mate had finished building his clinker before the voyage finished. With a huge sigh of relief I signed off in North Shields, England. I was August 31st, 1953. I got my travel warrant to go by train to Fishguard, where I boarded the Innisfallen for Cork.

1953; Radio Officer Jack Lynch on mv Goulistan

1953; mv Goulistan. My 9th ship

1953; Officers of the mv Goulistan in Oman, Persian Gulf

1953; mv Goulistan; Aground in Mauritius. Tugs assisting

Chapter Twenty-one

After a fortnight at home I was recalled to Liverpool, where I was assigned to my ninth ship, a passenger/cargo ship named m.v Goulistan. This time I had a Trainee Radio Officer named G. Robinson with me. The ship was owned by Stricks of London. Its Gross tonnage was 8430.35 and its Net tonnage was 5056.24. It carried general cargo but had a passenger quota of 12. Captain D.F.G. de Neumann was the Master in command. I signed on the Goulistan in Liverpool on October 15th, 1953. The voyage was to Mauritius, the Persian Gulf, North Africa, and back to the UK.

This trip was a new experience for me as I felt responsibility for twelve passengers, plus crew. Also, I had to train in a new Radio Officer who was on his first voyage. Despite the extra responsibility I looked forward to being on such a lovely ship. As I expected, the ship was spotless and well manned. My quarters were lovely; comfortable, beautifully decorated and they adjoined the Radio Room. For this reason, if none other, it was necessary to keep my area clean and neat, as passengers could come up to me to send messages home. My Trainee Radio Officer had an adjoining cabin. There were wooden decks and awnings for the passengers. None of my previous ships, with the exception of my first ship, the Isle of Guernsey, had wooden decks, which were scrubbed clean by the deckhands.

Everything ready we set off and sailed out into the Irish Sea, heading south. I allocated watch periods to my Trainee Radio Officer, who soon got over his nervousness so I let him perform his duties on his own. It was easier than I thought getting him familiar with the routine of watch keeping. I did occasional audits to check his log entries and maintenance records. At the beginning, with a few minor exceptions, they were fine. After we passed through the Irish Sea we again headed back to the open Atlantic where we proceeded towards Gibraltar and the Mediterranean and the Suez Canal.

During the voyage south a number of people handed me telegrams to transmit and I got to know a few of them. One lady passenger was from the North of Ireland and was going to the Persian Gulf to join her husband, who worked in Bahrain. She was quite a nice lady but I did not have much to do with her, or any of the passengers, as fraternising was not approved of for obvious reasons. When it suited this rule could be relaxed slightly, but under no circumstances would you be allowed to have a female passenger in your personal cabin. The Chief Mate explained that on a previous voyage a jealous husband came aboard to meet his wife, who was a passenger, and found her with another Chief Mate. The husband went bananas, took out a gun and fired a shot at the Mate. Fortunately he missed and the bullet lodged in the bulkhead in the cabin. The Captain, as a warning, purposely left that hole in the bulkhead to act as a deterrent to all crewmembers of the possible consequences of breaking the rules. This did not apply to the Radio Room however, where passengers frequently visited to telegraph messages home.

Having had a good crossing of the Bay of Biscay our trip continued through the Straits of Gibraltar, where now we were entering fine sunny weather with calm seas. My Trainee Radio Officer was enjoying life at sea and had settled in well. Onwards towards the Suez Canal, we passed North Africa on the starboard (right) and European countries on the port side (left). We eventually entered Port Said, Egypt to get our searchlight fitted in preparation for our trip through the Suez Canal. By this time my Trainee was thrilled with sea life and continued to wonder at the sights he was seeing. We joined the rest of the convoy and headed south through the canal. As usual we approached where the British soldiers lined the canal and gave the usual 'bird' to rusty ships flying the British Flag. This time, when we passed, spick and span and flying the Red Duster, there was a great cheer from the soldiers. They were proud to see a British ship as clean and well painted as we were.

At Suez the searchlight was removed and the pilot left us to proceed through the Red Sea for the Persian Gulf. Again, we had the usual pleasant trip, and the Trainee Radio Officer was agog with all the happenings, which to him were completely new experiences, whilst I had been through it all several times before, but was still enjoying it except for the thoughts I was having about the Persian Gulf. Whereas I had been trained on a Cross Channel passenger ship, on the same daily voyage, he was being trained on a deep-sea Passenger/Cargo ship going to various foreign places, and working various foreign Radio Stations, compared to my three British stations. He was enjoying himself and he didn't make bones about it.

Sometime during this part of the voyage, I noticed some unusual things begin to happen. Crewmembers began parading in similar new shoes, and I also noticed many of them were eating lots of chocolate. I was offered some bars, which I gratefully accepted, and soon got the shock of my life when the Chief Mate said that some of the cargo crates had been 'accidentally' broken and that the Captain was organising a cabin search for missing shoes, chocolate, and other materials. It was amazing to see how lots of this stuff started to appear floating in the sea, and very quickly the crew reverted to wearing their old gear again. Apparently this kind of thing went on lots of ships. It did not just stop with crew members. Dockers too were alleged to have been guilty.

We arrived at Ummsaid in Bahrain and some of the passengers disembarked. Here we unloaded our cargo and spent Christmas day which was a glorious sunny day, enjoying a sumptuous Christmas dinner and everybody was in great form. For once the wine and beer was out for this day and were much enjoyed. From here we sailed through the Gulf, and out into the Indian Ocean, where we then headed for Mauritius, off the east coast of Africa. We arrived in Port Louis in early January 1954.

While in port I played soccer for the ship's crew against the local police. Everything was going well until some guy tried a clumsy tackle on me and damaged the ligament of my knee. The doctor strapped it up, gave me painkillers and told me use a walking stick. Boy, did I need the walking stick! My knee gave me hell for what appeared ages, as I ran out of painkillers; I had no further treatment on the ship. It was agony.

On January 14th, 1954 a cyclone hit us while we were in the harbour and before we knew it the ship was up on the sandbank. Two tugs came to our assistance and got us off the bank and then we had to have divers go down to examine the hull of the ship for damage. It was exciting seeing all this happening and it took my knee trouble out of my mind for a while. Fortunately, the ship suffered no serious damage and was given the all clear. Some new passengers arrived prior to sailing and settled in for the trip back to the UK. I did not notice the passengers much as I was in my cabin resting my bruised knee, and ego. One day, while I was in the radio room, this beautiful dark-haired, tanned young girl appeared in the radio room, and smilingly asked me to send a message for her. As I turned in my chair I winced as I felt a stab of pain shoot through my knee. She immediately came towards me and asked what happened. I explained to her about the game and she offered to rub my knee. Decisions, decisions! Rules, regulations versus human nature! Fortunately the decision was not mine to make. A shadow appeared in the doorway and there appeared this tall lady looking perturbed.

"Yolanda" she called. "We must go now."

So this was the girl's name. Was this lady her mother? They certainly did not look anything alike. Hopefully, I would see her again. As it happened the lady was Yolanda's chaperone, and was to take good care of her during the voyage as she was apparently the only child of a wealthy family, and was going to England for educational purposes. We used to see each other on deck, and chat a lot under the watchful eye of the chaperone who was never far away. The chaperone mellowed too and eventually all three of us became good friends.

As we continued our voyage back we called in at Aden in the Yemen, at the entry to the Red Sea. Many ships call here to get bunkers on their journeys as the cost may be lower than elsewhere. It is a great port for bargains. Bargaining and bartering is part of the fun. Cigarettes were always a good bartering item and we got them cheap onboard. They were tax-free and usually could be used in any lesser-developed country as cash. In exchange for a lovely silk shirt, I handed over a carton of Lucky Strikes and was delighted with my bargain, which I thought was very cheap. These guys will bargain but this time I got the worse end of the deal because after we left port and I opened the shirt, I found a large mark on the front of it. However, I decided that I was not to be the fall guy. I repacked the shirt neatly, to barter it at our next port of call.

We kept on moving and the ship went through the Canal for the umpteenth time as we headed east towards Ceuta, in North Africa. We tied up there for cargo and this is where I saw shocking sights of girls at their lowest. We were only tying up the ship when local prostitutes started to board. These prostitutes are not welcome onboard but they sneaked aboard before they were spotted. The Mate did not even have time to put watch-keepers on the gangway before one passed me. I nearly got sick just at the sight of her. Her eyes were streaming and were blood red. Her nose was running and there was a smell from her that would turn your stomach. How fellows go with girls like this, I asked myself, was a mystery. I don't know how many got aboard but they vanished aft into cabins and eventually started to come amidships. When they were found by the security they were bundled off the ship. I happened to spot my Trainee Radio Officer being chatted up outside his cabin and was embarrassed at this sight. I advised him to get rid of her and warned him of the dangers. I also told this woman to get away from the Radio room area.

As well as these poor girls some of the usual 'salesmen' came aboard to barter. Lots of tapestries and brightly coloured cushion covers and lengths of suit material were readily available. I saw one guy who had perfumes for sale and I bartered the stained shirt for a bottle of Chanel No 5. This to me was a good bargain, and I went away happy in the knowledge that I had not lost out on the deal as the perfume was genuine.

I was glad when we cast off again and headed back home. However, we had one more call to make, and that was to Casablanca. Ashore here it was fun to see the girls behind their veils, a new sight for us all. Some dropped the veil slightly and peeked out and gave us a big smile. It was so different from Ceuta. Our stay here was very short and soon we were off again to take on the pilot and head out to sea.

On our way to the UK we proceeded back through the Bay of Biscay, up the English Channel, up the North Sea, and back to Hull, where we arrived on 18th February, 1954. I reported to the Marconi office, gave my Trainee Radio Officer a good reference, and said goodbye to him. I also bid farewell to Yolanda and her chaperone. Unfortunately, none of these people ever crossed my path again.

I signed off on February 19th, 1954 and again I got leave for about 10 days. I gave the bottle of Chanel No. 5 to Molly and she loved it. We did the usual visits to the cinema and drove around to various places, including Blarney Castle where we both kissed the Blarney Stone. At home mammy and daddy and my sisters asked lots of questions about the trip and I think daddy envied me the experiences of travel that I was having. The ten days flew by and I returned to Hull where I was reassigned back to the same ship on my birthday; March 5th, 1954. I stayed onboard until we set out for Liverpool, where we arrived on the March 11th, 1954. I signed off and left the ship. There was no Trainee Radio Officer for this short trip to Liverpool and we did not have passengers. I was sorry leaving this ship as it was one of the best ships I had been on to date but I was looking forward to the next adventure.

While in the Liverpool Marconi depot I was called to the office one day and a smiling manager took me aside and said they were offering me a unique job that would result in the top rate of pay available for a Radio Officer. My ears pricked up at this but at the same time I withheld my excitement as I had never heard of the Marconi Company doing anything for a Radio Officer unless there was something in it for them. I sat down and was amazed when this guy said the job was manning a Radio beacon on an island in the Persian Gulf. I would be there on my own and for an undisclosed period of time. I had experience of the Gulf before and it was not on my first choice of holiday resort so it did not take me long to give my answer to the manager. He tried all his persuasive skills to change my mind but there was no way I was going out there. No thanks!

    My decision made, I stayed on standby in the Liverpool office and within a few days I was sent to join my next ship, hoping and praying that it would not be to the Gulf.

Chapter Twenty-two

My tenth ship was the Tramp steamer s.s Selector. It was my ninth deep sea voyage and was to the Leeward and Windward Islands in the Caribbean Sea, British Honduras, in Central America and back to the UK. The ship was another Liberty type ship and was registered as No.169505 with a Net Tonnage of 4743. The Master of this ship was Captain R.L. Williams. When I signed on in Liverpool on March 13th, 1954 I was not aware of the fact that this would be the last British flag ship that I would sign on for a long time.

When the day came to board I took a taxi to the docks at Liverpool and found the Selector tied up, and looking like any other merchant ship. A dark hull and white superstructure greeted me. There was nothing exciting about the exterior of this ship so I was quite surprised when I went aboard to find the Captain in full uniform, including peak cap. Normally, the Captain's would dress less formally so this made me wonder who this man was. I soon found out when I went to sign on. He told me that he was a Royal Naval Voluntary Reserve (RNVR). This meant that he had either served as a fully commissioned Royal Naval Officer and retired, or that he had volunteered for active duty if he was needed. I never bothered to find out whether he had seen active duty during the war. So now I was going naval with all this entailed. We would have to be spick and span and in uniform at all times. This did not sit easily with me, as I liked to dress in Khakis, or anything I felt like, except when I was on a passenger ship. However, I had to wait and see how things would work out on this ship before I took any chances. As it happened, I very nearly did not sign on this ship, as I had a run in with the Captain at the signing on process. He was filling in the data for me when it came to the question of Nationality and he said; "British."

I answered, "I beg your pardon but I'm Irish."

He immediately repeated "British" and I repeated, "Irish."

He then started to get hot under the collar and said, "You were born in 1927, which makes you British."

I said, "I was born in Ireland in 1927. I am Irish and proud of it. I have signed on all the other British ships as Irish and I don't intend to deny my nationality to make it convenient for you because your crew list would show a foreigner amongst the British crew."

He then got mad as hell and said, "The 'Black and Tans' were the boys to sort you lot out."

By now I was fed up with this guy and I replied, "Your murdering friends should never have been let loose on innocent people and if they were your friends, which you admit to, then I want no more to do with you, or them, Goodbye. I'm not signing on this ship."

I picked up my bags and started to go down the gangway when he came after me and apologised. He asked me to reconsider. I think he knew he had overstepped his authority and seemed sincere. He offered his hand in a handshake and I took it.

I said, "Captain, if you want me to sign on, it will be as Irish, and I don't want any remarks or baiting about politics, or religion, during the voyage. If you agree to this I will sign on."

He accepted, and he kept his word for the entire voyage. In fact we became good friends, and he told me later that he admired how I defended my right to be Irish.

After the first period of the voyage which was in the Home Trade I signed off in Liverpool on April 2nd, 1954 for a week's leave only to return to rejoin the Selector and prepare for the next part of our deep sea voyage on April 14th, 1954.

On board we had a Scottish Chief Officer who used to listen to his Eddystone receiver (Short wave radio) whilst he was off watch. It was his favourite pastime. One night the Captain came to my cabin, during my time off, and asked me to send an urgent telegram for him. The Chief Mate was also off duty, listening to his radio as usual, when I started up my transmitter and commenced sending the Captain's message. All hell broke loose as the Chief Mate came storming into the radio room and ordered me to shut down, as I was supposed to be off watch and was interfering with his radio. I told him to leave the radio room and mind his own business. Just then the Captain came rushing into the radio room, to find out what was going on. It did not take him long to get the drift of things and he told the Chief Mate in no uncertain terms to leave the radio room, and that he had no jurisdiction over me or when I worked the radio. The Chief Mate avoided me for the rest of the voyage and I shed no tears for him.

On this particular voyage we had various items of cargo, and amongst the cargo was a deck cargo of long steel pipes about ten inches in diameter. There was also one car on deck. This car was for a special customer in the island of St. Lucia in the Caribbean Sea. The ship and cargo was heading for the Caribbean Islands where it was being delivered to various customers in various destinations in the Islands.

Along the route on this journey we were lucky enough to island hop to a lot of the nicer Ports and Islands. In the Leeward Islands we visited Basseterre in St.Kitts, Cul de Sac bay in Dominica and to St. Lucia in the Winward Islands. We also tied up at Montserrat and Martinique islands, nowadays exotic holiday destinations though in those days they were very underdeveloped. We berthed in San Juan in the Dominican Republic and in Kingstown, St. Vincent and finally St. John in Antigua before we crossed to Belize in British Honduras, Central America, and Houston, Texas.

When we reached the Caribbean we discharged the pipes, plus other parts of the general cargoes in various ports. We called to Belize, for bunkers before setting off for Houston, Texas. Thankfully each trip was short and within sight of land so we were never more than a few days from port. I got ashore in most ports and enjoyed the quietness and calm that prevailed on the islands. The local people were friendly and smiled a lot and there was no sign of begging. Children ran around happily playing and loved it when I spoke to them. Sadly, it was years later when I found out about the close association between the island of Montserrat and Ireland. Apparently there are many locals there who have Irish names! I only heard of this connection when the local volcano which overlooked the town erupted and almost obliterated the town. In 1493 the island was discovered by Christopher Columbus and the British colonised it in 1632. 20 years later a large contingent of Irish settlers moved there from the neighbouring island of St. Kitts. Monserrat is also known as the Emerald Isle because of its lush green tropical landscape.

I really enjoyed the beautiful sights on the islands, they were stunning. I loved to watch sunsets, like nothing I had ever seen before, drink in the amazing colours of the various flowers and trees, and listen to the wonderful sounds of exotic, multi-coloured birds whistling and flying close by. Local fishermen would come out and try to sell us fresh fish and we bought some lovely lobsters which made a wonderful change in our menu.

After all this peace and tranquillity it wasn't long before trouble came our way again. We were in Houston, loading some cargo of grain to take back to the UK when disaster struck. As we manoeuvred to get alongside the docks with the assistance of two tugs, one fore and one aft I was on deck watching when the tug at the bow began to push and nudge the ship and the tug astern awaited the order to move in to assist. The tug was about seventy feet off when she got the pilot's signal to push the stern. I couldn't believe what happened next. Amid screams and shouts from the pilot and Captain, the tug came towards us at full speed and caught us in one hell of a wallop astern, rocking the ship massively and jolting us all from our stances. When all the commotion died down and everything was stabilised a brief investigation showed that the Mexican Captain of the tug was drunk. Inspection of our ship showed that some metal plates were buckled and some rivets popped so it was decided that we would do temporary repairs until we got back to the UK when the permanent repairs mending could take place. On the inside of the ship a wooden frame was built around the damaged area and was then filled with concrete. The temporary repairs complete and with all the crew crossing their fingers they would hold, we left port and headed back across the Atlantic with a load of grain. Very mindful of hurricanes because this was the season, we hoped we would get away and avoid any that might evolve. Though we didn't get caught in any full blown hurricanes we did hit some bad weather and rocked and rolled as we danced our way home over leaping waves. Thankfully, and to our great relief, the temporary repairs held up and we arrived back intact and much relieved.

    Finally I signed off in Manchester on July 7th, 1954, and I took a train to Liverpool and reported to the Marconi office where I handed in all my documents and log reports. Whilst sitting around in the office waiting I got fed up and made the decision to try working ashore for a change. After a few days I heard about a job ashore and I decided to 'Swallow the Hook' (Give up the sea) and work on land. I went into the office and in one of my biggest decisions to date, resigned.

    "You'll be back," the clerk said.

    "Not me, this is the end of my sea going days." I retorted, and I meant it at the time. It was July 10th, 1954. This was going to be a momentous time for me and I had taken the first step of settling ashore....

Chapter Twenty-three

I had now retired from sea life and was wondering if I had made the right decision. For better or worse I had to try a shore job and I reasoned with myself that I could still go back to sea if I felt like it. So off I went in July 1954 for my first interview and hoped that all would go well. I didn't like the idea of going back into the Marconi office begging for my job back so I was determined to make things work. The job I was applying for was working as a Radio operator with the British Civil Service at the American Air Force base in Burtonwood, Lancashire, just outside Liverpool.

After a tough interview I was delighted to hear that my application was successful so I embarked on the next part of my life; I was attached to the United States Air force, under the 'Native Son' programme. This apparently was the name given to any non U.S. person who worked for them when employed by the British Civil Service. I worked in a radio room at the USAF Airforce base, and had a fellow Irishman from Limerick with me for a while in the same capacity. He too was an Ex-Radio Officer but he left soon after I got there. Our job was to maintain constant radio contact by Morse with all the US bases in the UK and some in Europe. I remember the call sign of this station was AJC20 As this was the biggest United States Air Base in the UK in WW II I was kept busy taking stores lists and other messages which were confidential under the secrecy laws. I had to sign and was sworn to uphold these secrecy laws governing all messages. This was a USAF maintenance base for C-54s used during the Berlin Airlift. During WW II this base was very busy and many planes operated out of here. Whilst I was there lots of planes kept coming and going and it housed a lot of scrap planes as the years passed.

The same daily travel and being alone in the radio cabin soon made me wonder if I had made a big mistake giving up the sea life. At least there I had plenty of company and split watches which gave me time to mingle and have a bit of craic.

I soon found the job tedious and boring and with no chances of promotion I decided to resign, and look for a better job with more prospects elsewhere. The money I had been earning was mediocre enough so I didn't feel too worried when I tendered my resignation. When my boss heard I was resigning I was immediately called into Major Strong's office for interview. This was my top boss in charge of the Radio shack.

When I got into the office I got a shock because I did not expect Major Strong to be a woman. She was a well-built lady, with a nice manner, and when I told her I was going for more money she said she understood but could do nothing about that as the British Civil Service controlled pay scales. She said she would willingly give me American rates for the job, which were far better, but she could not override the existing structure of payment. She then suggested that I might consider joining the American Airforce, and she would be glad to approve me for acceptance. I had not considered this possibility and said I needed time to think about it. After some thought I decided to start afresh and leave altogether, thanking Major Strong for her offer but declining it nonetheless, I had worked here for seven months. I said farewell to the lads I knew, and to Major Strong, and returned home to consider my next move. There was no chance of getting work at home and to be truthful I still missed the sea life.

It was now late January 1955 and with time on my hands I enjoyed being at home again and Molly and I had a great time. My parents were delighted to see me and were quick to tell me all the stories of what had been going on in my absence. They related a funny episode that had taken place in one of their ventures. Daddy and mammy used to let out rooms to some couples from time to time and told me about one couple who had rented out two rooms upstairs. The couple, Billy and Nellie, got on like a house on fire except when Billy arrived home after a hard day's work. It was in the middle of winter and very cold. As he got in to his room shivering, Nellie was asleep and well covered up. Billy shouted, "Fire Fire," Nellie awoke startled and sat up in bed shouting, "where's the fire?" Billy shouted back;

"Exactly, where is the fire? It's in every F****** house in town except here!" He then threw a jar of water over her in the bed. The argument went on for ages between them and mammy and daddy were in hysterics recounting the tale.

Now that I was back home and with some money of my own I was pleased to be able to give some to mammy. Daddy went and bought a Hoover vacuum cleaner with this money. He seemed oblivious to the fact that we did not have one carpet in the entire house. All the rooms were covered in Linoleum or the floors were painted. He needed a vacuum cleaner like he needed a hole in the head. He claimed his reason for buying this machine was because poor Billy, who lodged upstairs, was a salesman for these vacuum machines and daddy thought that Billy could do with the commission. So in fact I supplemented Billy's rent to pay daddy. The logic of what he had done escaped me. My sister later told me daddy used the machine to vacuum up plaster which had fallen down after he had being doing a job in one of the rooms so I guess it came in useful in one way.

Having returned to Liverpool after two weeks, and after some more consideration, I decided to try freelancing as a Radio Officer because the money was better than that offered by Marconi. I headed for the Radio Officers' Union offices, and for the first time in my life I became a Union member. They soon offered me the chance to start a new and rewarding career as a freelance Radio Officer. On March 19th, 1955 I joined my first Liberian Flag ship in my new role as a freelance Radio Officer, and other added responsibilities

Chapter Twenty-four

My first freelance ship was my eleventh ship to serve on and was a Liberty tanker, named s.s Albert G. Brown. She was registered in Monrovia, Liberia. Her Gross tonnage was 7289 and her Net tonnage was 4475. Her official No. was 406. Onboard she was always referred to as the AGB. Being a freelance Radio Officer I was now in a position to negotiate my own wages and terms of employment within the regulations governing seamen and their terms of employment. I would be a shipping company employee now and not a Marconi man.

The owners of s.s. Albert G. Brown were Bernuth Lembcke Company, New York, also known as Torres Shipping Company. When I joined they owned two tankers but later purchased a third tanker. The AGB was commanded by Captain Julius Vice and there was also a Relief Captain, named G. Catlender.

Unfortunately, I encountered a problem when I first met Captain Vice. Initially, I seemed to unintentionally aggravate some Captains, I wasn't sure why. We were discussing the terms of my contract when the matter of salary came up. Captain Vice asked me what I expected and I said the minimum Union rate. This was about £60.00 per month. The mention of Union, however, nearly made the Captain choke on his Lucky Strike cigarette.

"We don't have union people on this ship," all six feet of him drawled in his Louisiana voice. He continued, "I will offer you better rates, but not if you are a union member."

"How much will you give?" I asked.

We struck a deal on a salary, and so my union membership was bought out. I had no regrets about this because I had never been a union member, though I never told him that. So, with everything settled and my terms agreed I was off to sea again.

The ship was waiting in Stanlow, Ellesemere Port on the Manchester Ship Canal and I joined her there, feeling good about my new role and happy. I signed on this time on the 19th March 1955.

My tenth deep sea voyage had included trips to the East and West coasts of the USA, Cuba, Aruba, Curacao, Venezuela, Panama Canal, Puerto Rico, France, Germany, Holland, and U.K. This ship differed from normal tankers in as much as it was exactly the same as a Liberty cargo ship, but had tanks instead of cargo holds.

Little did I think when I joined this company how much leave of absence I would get, due to ships being laid up over the years, and due to my illness while I was with the company. Torres Shipping Co. turned out to be great employers and I really enjoyed my time at sea with them.

Agro was to rear its ugly head again! On my first day aboard this ship I met the American Chief Mate, and the Third Mate, who hailed from the Cayman Islands. I immediately took a dislike to the Chief Mate who appeared to be drunk. He had thin lips and watery eyes that tried to size up my reaction to his taunting. The Third Mate, Colby Jackson, was young and had a wholesome smile and relaxed demeanour. I liked Colby, who was later promoted to Second Mate, and we got on very well during our time together. However, my first run in with the Chief Mate was on this first day.

He greeted me with a snide, "Hello Rodney," in a mock English accent. He was not even aware of my first name at the time and I could sense he did not like English foreigners, and I was classed in this category.

I looked him straight in the eye and said, "My name is Jack and don't forget it. I'm Irish, not English and I don't take kindly to people who ignorantly try and pick arguments or fights." My blood was beginning to boil.

I watched his eyes while I said this and saw flicker of doubt creep in to him. At the same time I shook hands with

Colby, who was taking it all in. Later, he told me that the Mate was "like that" and to ignore him.

I said, "No I won't, and I will not take any snide remarks from him." I never had a problem again with this fellow.

All the deck crew and some engine room lads were from the Cayman Islands. The deck officers were American, as were the Engineers. I had other duties onboard as well as Radio; where I had to liaise with immigration, shore Doctor and Customs' officers as well as act as Medico on board. Since I was now a Company employee I also had to assist with signing on/off crew members and help the Captain run the 'slop chest' on a commission basis.

The 'slop chest' was a store or shop where we had all kinds of stuff for sale. I got plenty of experience with all these jobs during the voyages and I enjoyed it a lot. I had a number of ways of making money onboard the ship. There was the 'slop chest,' which I opened a number of times a week and the crew purchased cigarettes, Nivea cream, Khaki clothes, chocolates and various other bits and pieces. I was a glorified shopkeeper. I could not get over the fact that the black lads were using Nivea to protect against sunburn. It did not dawn on me that they too could get sunburned. The Captain bought the stock for the slop chest and I got 50% of the profits from sales for running the shop. It turned out to be a good money spinner.

During my radio duties I used to get the crew to send flowers, or candy via Interflora to their wives, and girlfriends, and I got commission from McKay Radio for each sale. I was always interested in making money legitimately so I decided to chance my arm by asking for a salary increase.

On one trip I told the Captain I was going to resign in New York, as I could get more money from National Bulk carriers, freelancing with them. As it happened the ship's Superintendent was due to come aboard there, and he immediately said he could not increase my direct salary, but that I could claim an extra $100.00 U.S a month in overtime, each and every month if I stayed. I did! The overtime went against battery and equipment maintenance. In fact, I had hauled off a big bluff, because I did not really intend to resign from this company, and at the same time I was sick in the stomach wondering what I would do, if the Captain decided to accept my resignation. Luckily, he didn't call my bluff.

The Captain, who was from Louisiana in the United States, also ran a farm at home as well as being a sea captain. He stood about six feet two inches tall, and had a loud Southern drawl. He once told me that on his farm he had a big bull, and whenever he got home he would go to the gate and give a big bellow, and the bull used to reply to him. He was very proud of this. Generally, he was one of the boys, but on occasions he could make you understand who was in charge. He was tight with his money, and being a smoker tended to 'borrow' cigarettes from crew members, but I never remember him handing any of his own around. Usually, he smoked Lucky Strikes, in a long cigarette holder.

His uniform was a Khaki baseball cap, with khaki shirt and trousers. He was on the bridge quite a lot; I believe this was for company more than anything else. We got on very well, and became good friends. He told me all about his family and I met his lovely wife twice. She used to call me 'Dear John' in a nice friendly way. She was dark-haired, slim and full of life. She was a great woman who later died in her prime. I think it was from cancer. The Captain was devastated. On one occasion that lady drove her car from Louisiana, to Portland, Oregon to be with her husband when we hit port. She told me she had driven in her bare feet, as she found it more comfortable.

Captain Vice told me once that he had risen from ordinary seaman during the war, to the rank of Captain in a short time. This was due to the shortage of officers to man the large output of Liberty ships. He said that they were sent to convoys and told, 'Follow the ship in front,' not knowing if the Captain in front was a rookie, and first-timer like himself, or an experienced seagoing Captain. It was hair-raising to say the least, and the biggest fear he experienced was losing sight of the convoy and finding himself alone in the Atlantic Ocean with the German U-boats scouring the

Atlantic for Allied shipping. He said he soon became seasoned, and thankfully he survived without any mishaps.

Up to joining this ship I had been used to eating an Irish or English dinner of potatoes, meat and vegetables etc. I was in for a shock on board the AGB though. As the Captain had a farm he believed that by introducing and using the same products he grew on his farm at home would help keep the prices up and one of his products was rice. That was bad news for me! Rice and Grits became the main dishes on board and for an Irishman like myself, who had associated rice with dessert, and not as a main meal, the lack of potatoes at dinner times was an area of contention between us. None of the rest of the crew appeared to be too worried about the infrequency of spuds on the menu but it was a sore point for me. However, we both compromised and I took rice but also got potatoes every so often. The Captain bitched because he said he was there to promote the sale of rice. Eventually, I got to like rice as a main dish when I opened my mind, and I also got used to chilli and Tabasco dishes. I remember the shore Superintendent asking the Captain for a fan to sit on after eating some of the spicy hot dishes!

Captain Vice had no inhibitions about saying things which might upset people, and which he thought were funny. For instance, my radio room was close to the bridge and sometimes, when off watch, I would saunter in there and listen to the BBC world news and music. On one occasion, we were steaming along with one of the black crewmen on the wheel. This fellow had a good sense of humour, and it's just as well because the Captain lit a cigarette and drawled, "Connors, you're not just black, man, you're shining."

I felt for Connors, but he just looked over his shoulder, and with a perfect set of sparkling white teeth, grinned at the Captain and carried on steering the ship. Poor Connors on one occasion came aboard drunk out of his mind and started a fight. It really got out of hand, and the Captain tried without success to stop it. He asked me to accompany him to his cabin, where he opened a safe and took out handcuffs and a revolver. I got really nervous as I had never encountered anything like this before. "I will stop this mutiny," he said, agitatedly. I liked Connors and felt that this was just a blip, a one-off by the guy. However, when we got to the cabin Connors was lying down but jumped up when he saw the handcuffs and started to scream, and flail his arms. Two other deck officers subdued Connors, and handcuffed him to the bunk, which was secured to the steel deck. We all departed and locked the cabin door behind us. There was a hell of a rumpus going on inside the cabin and eventually there was a crashing noise as Connors moved the bunk, and bashed it at the door. He did not succeed in breaking out and eventually slept off his stupor. His cabin was wrecked, and he felt so humble and contrite when he came back to his senses. The Captain fined him the cost of the damage but let him stay onboard.

Another fellow, a Dutchman, came aboard drunk another day and started fighting. The Captain warned him that he would cuff him if he did not stop. He did not and the Captain again got out the handcuffs and asked me to accompany him. The Dutchman was slightly built but wiry, and the Captain had difficulty trying to secure him. I spoke to the Dutchman, and asked him, for his own sake, to surrender to the Captain, or the consequences could be severe for him. He immediately held out his hands, and the Captain put one cuff on his hand, and the other end of the cuff was locked onto the mast down stay. The Captain then left, and the Dutchman said to me; "Sparks, the jails in Holland are better." He was let loose about four hours later and discharged from the ship at the next port.

There were a number of humorous crew escapades, and tragedies, during the various voyages on board the AGB. Life on board was never dull, that's for sure. I remember the time the Captain showed me a new Leica camera that he had purchased for his trip to London to, as he put it, 'photograph all the wild animals in Piccadilly Circus'. You should have seen his face when I explained the facts to him. Priceless! I wish I could have photographed him then!

Many of the crew from the US found London to be quite a new and exotic experience when we were stopped there for ship's business. The elderly American Chief Engineer had his first outing on the top deck of a double-decker in London. He was petrified when he saw the oncoming traffic on, what he thought, was the wrong side of the road. He ducked down in his seat and covered his eyes, shrieking in terror. I'm sure he must have needed a change of undies after that! The rest of us who were not so naïve were in hysterics! After this experience he refused to go on any bus in London again.

Again, on a stopover in the UK, I had a problem in Liverpool with Customs when they came onboard to search the ship. I was quite calm, used to the experience of Customs checks, and did not expect any problems when the Customs Official came into the radio room. He asked me the usual question;

"Have you anything to declare?"

"No." I replied, without hesitation.

He started his search. He eventually came to the cabinet that was used for spare parts, and went though each shelf. Eventually there was only the kick panel left, and he took a metal lever known as a jemmy out from his bag, and levered the panel away. He then stooped, put his hand into the open area, and pulled out a carton of Chesterfield cigarettes, asking me if I knew anything about the carton. I said that I had no knowledge about the cigarettes. He then pulled out another carton of Chesterfields, and again asked if I knew anything about them, to which I again denied knowledge. My stomach at this point was churning.

"I only smoke Lucky Strike, and I can prove it." I thought to myself. These thoughts were rudely interrupted by his voice; "Or these or these?"

Looking at him I saw he was now holding two more cartons, but my confidence in my innocence was shot when I saw the brand, Lucky Strike. Another carton was taken out and the Captain was called to the Radio Room. He looked at me and asked me about the cigarettes, and I told him I knew nothing about them. I also pointed out that there were very old looking marks and stains on the cartons, indicating age. Also, I noticed while the customs man was removing the cartons, two of them appeared to be stuck to the deck indicating years of storage. When the senior customs officer found out I had been on the ship for only a few months he immediately said there was no way that I could have been involved, on account of the state of the cartons. He then advised me that when I join a ship I am responsible for my cabin and the Radio Room, so I should do a search of all areas in my rooms. He said I would be responsible even if stuff was found behind panels. Frightening to think how easily you can get blamed for somebody else being thoughtless.

Thoughtlessness by one of the crew nearly drove me mental for some weeks. We had a big black guy on board who started waking me up anytime between two a.m. and six a.m. complaining of a toothache. He had a large black hole in a molar. The usual treatment was to use cotton wool, soaked in oil of cloves, to dull the pain. This I kept doing, and told him to go ashore and have the tooth attended to at our next port of call. He refused out of fear, so I decided to frighten the life out of him anyway the next time he woke me up with a toothache. To this end I went to the medical locker and got out a tray, and on this I placed the biggest hypodermic needle, a bottle of Novocain, a large tooth extractor, cotton wool, and a large napkin. I covered the lot with a white towel and had it ready for my patient. It was not long before I got the wake up call. I immediately got my tormentor to sit down, and then went through the formality for setting my man up. I produced my tray of implements and slowly removed the towel. The patient looked at it warily, absorbing the contents. I picked up the large needle. "Sparks, what you going to do?" He half shouted in a scared voice. "Open your mouth wide until I get room to inject the Novocain, to take out this bad tooth." I said.

The whites of his eyes nearly blinded me. He jumped out of the chair, shouted that the pain was gone and then, so was he. I had no more disturbed nights after that.

When we hit port the doctor used to come onboard to do the normal health checks. As medico it was my job to sit next to the doctor, and with the crew list identify each sailor for the doctor.

We had a small little Cuban man named Hernandez onboard who worked as a greaser in the engine room. Poor Hernandez had bad haemorrhoids. The doctor said he wanted him in hospital for examination and asked me to prepare an enema that I was to give to Hernandez first thing the next day. I had never given one of these before and so the doctor explained the procedure to me so that Hernandez would be 'cleared out' before his visit for examination. Early next morning Hernandez came to the sick bay, which was a cabin used for medical purposes. As he entered the door, there was a bed straight in front on the opposite bulkhead (wall). To the right was the toilet, about six feet from the end of the bed. There was a secure locker, with all the requirements for medical assistance, and I held the key to the cabin and locker. I told Hernandez to lie on the bed facing the bulkhead with his trousers down and his knees drawn up under his chin. I then applied Vaseline and inserted the nozzle while I held the container with warm soapy water up above his body. As the container started to empty poor Hernandez became uneasy. "Sparks, I burst, I must go," he cried out in pain.

I could see that he was not kidding and I pulled out the tube and headed straight for the door. Poor Hernandez twisted off the bed, got to the floor, and made for the loo. Alas, he never quite made it, and there was strange sound as the enema, and other foreign parts littered the floor. The stench was overpowering and putrid. I was glad I did not have to clean up the mess. Later, when Hernandez cleaned himself and had a shower, I escorted him to the hospital. His examination was not very promising, and he was discharged from the ship and sent home for treatment. I never found out what happened to this cheerful and nice man.

Whilst with the AGB we went to Galveston, Texas on a few trips. Colby Jackson and I were ashore one day, lying on the beautiful beach, watching the beauties parading and swimming in front of us. We had our sunglasses on and handkerchiefs over our heads to protect us from the blazing sun. Right in front of us were three young teenagers about seventeen years old. They had lovely figures decked in bikinis. Close by us was a middle-aged woman, who was apparently a chaperon or their mother. She kept calling the girls, and we were slyly watching them because the woman was possibly watching us, watching them. Eventually, the woman called the girls ashore, and they ran out of the sea and started to take off the plastic lifebelts as they came out. Two dropped the belts and stepped out of them, but the third girl struggled to take it off over her head. She did not notice that as she did so she lifted the top of her bikini off her breasts. Merrily she kept running towards the woman, who was having hysterics trying to warn her about her bare breasts. By now Colby and I were sitting up, with sunglasses removed, having a grandstand view. That is until the woman spoiled it by running between the young girl and us blocking our lovely vision! The hen protecting the chicks!

Most of the crew were acting responsibly and behaving properly but I remember a Norwegian Second Mate got into serious trouble here in Galveston. He liked the booze, and whilst ashore he started boasting about communism and degrading the USA. That was it. He was arrested and ordered to leave the country never to return. He was placed under armed escort and returned to the ship, after a brief court appearance, where he was confined to his cabin. He tried to bribe me into getting him a bottle of Coke, spiked with vodka. He told me he would give me his expensive Omega watch in return. I refused as I did not want to cross the US laws. An armed Pinkerton agent was stationed outside the cabin for the duration of the ship's stay. Also on the gangway was a sheriff, who had a large .45 colt revolver on his waist. He spent the time whittling away at a piece of wood when he was not chatting to someone. I asked him one day why he was there as well as the Pinkerton man, and he smiled as he told me about his friend Bill. He said I would like Bill, who was a kind, gentle and very nice man. Apparently, Bill was on duty like our friend, and was securing against a Greek, who was classed as undesirable and was awaiting deportation. One night, Bill heard a splash in the water and saw the Greek making a break for it. Bill pulled his gun, and when the Greek ignored the call to surrender Bill let go with a shot just over the Greek's head. Still the Greek continued to swim.

Bill shouted, "The next bullet goes through your head, boy, so turn round and come ashore." The Greek obeyed, and Bill arrested him.

"You'd like Bill, he's a nice fellow, kind and gentle," said my sheriff storyteller.

We continued to swap molasses and Crude oil whilst travelling between ports in the Caribbean and the U.S. and eventually loaded a cargo of oil for Stanlow, in Ellesmere Port.

On the trip back to the U.K I sent flowers to Molly and hoped I would not get into trouble again like the last time I had sent flowers. I had previously sent flowers to her, for her birthday, and when I arrived home some time later I was met with a frosty reception from her. She accused me of being home at the time of her birthday and not going to see her. It took me some time to convince her that I had not been home when the flowers arrived to her but that I had used Interflora. She wondered why the flowers had come from the shop in Cobh and not the shop in Cork, closer to her and was convinced that I was not telling the truth that I had been home in Cobh and had not gone to visit her. It took me some time to convince her otherwise but eventually she believed me. She told me the flowers were lovely, but her thoughts about me being home had spoiled the impact. What I wondered did she think I was up to? I never did find out, even after we got married.

Once, during the trip when I contacted Valencia Radio (EJK) to send a Ship's letter telegram message to Liverpool, the Radio Operator at Valencia advised me to try and send it direct to Landsend Radio (GLD) or Seaforth Radio (GLV) or some other station in the UK, as it would take some time to get from Valencia because they only had 'Pony Express' there. A ship's letter telegram is a communication where a message is sent by radio to the Coast station and is posted by mail to its destination from the station. This was a cheaper type of message to send. It sounded quite funny, but I had actually been trying various UK stations, on medium wave, without success, before trying Valencia. However, I went on to short-wave radio and sent the message through Portishead Radio (GKL/GKS/GKN).

We continued up the Irish Sea and berthed at Stanlow, Ellesmere Port. This was to be the last time I would ever dock at this port as the oil crisis was looming and cargoes became sparse and difficult to obtain. I signed off in Manchester on 7th February 1956 and returned home on full pay while the ship was laid up.

I was greeted happily by all at home and this time the flowers had been delivered to Molly by the Cork shop so I got my brownie points. During the stay Molly and I met quite often and I used to drive around quite a bit and go out visiting a lot but was getting bored and was eager to get back to sea. I had to go to Dover to join my ship again and re-signed on five months later, on the Albert G. Brown in Dover, England on 17th July, 1956.

Most of our trips from now on tended to be with cargoes of molasses from Cuba to Jacksonville, Florida but not every trip could be guaranteed. After any voyage when we transported crude oil, or any non-edible fluids, it was necessary to clean tanks. Usually, the cargo to follow oil was molasses, or fresh water. The cleaning was done using high-pressure hoses that had triple swivelling nozzles with hot water. It was necessary to be in the tank with the hose when this operation was in progress. Most of this cleaning occurred in the Caribbean, around Cuba, where we loaded the molasses. This could often be a hazardous job.

On most trips the ambient temperature was very hot, and I got good rates of pay helping out. On one occasion I was down in the bowels of the tank, using the hose, and after a few hours I felt pain in my chest. It was a bone type pain, but I also felt very weak. I had not drunk any water so I was feeling dehydrated. I decided to climb up the ladder to get out. I don't know how I managed to get up with the weakness and pain, but at last I got my head above the deck and staggered out. The Captain gave me salt tablets and told me to drink fluids and rest. The doctor diagnosed a cracked clavicle and this finished my tank cleaning for the rest of the time.

Another accident happened to my friend Colby when he was cleaning the tanks. The hose jet skimmed across his eye and caused temporary sight loss. He was lucky that the jet only skimmed across his eye, and even though painful, there was no permanent damage, thankfully.

Our visits to Cuba were frequent, as we picked up Molasses for the States. Vita was our principal port of call, and sometimes we called to Newvitas, which was close by. At the time, Fidel Castro was in the hills fighting the soldiers of the Baptista regime. I heard constant shooting up the hills, and watched as soldiers and police patrolled the streets. There were sandbags everywhere protecting them and it was an uncomfortable sight.

On one trip, as in all ports, the conmen were around trying to fleece us seaman. I was in my cabin when a fellow came in, and looked around as if he suspected someone was within earshot. When he was happy that all was well, he sidled up to me offering me a, 'genuine diamond ring' which he had stolen from a wealthy house. It was cut price but definitely genuine he assured me, and he scratched the porthole glass with one of the diamonds to prove it. Eventually, I succumbed to the temptation and gave him two cartons of cigarettes for it and he then scrammed. I kept looking at the ring and felt guilty, and also was not confident that the ring was gold, or that these were real diamonds. After a day I decided to take a chance, and go to the police with the ring, hoping that I would not be classed as a criminal. The police station was close by and was heavily guarded and protected by sandbags. To cut a long story short, the police looked at the ring once and laughed heartily, telling me you could buy them in the local market for a few cents each. I was not charged and went back to the ship feeling silly, and vowing never to be caught again. I still have this ring though!

One night, there was panic ashore as Castro's men came down from the hills and attacked the town. The crack of rifles and machine guns sounded very close to us, and I saw the flashes as bullets were fired. Baptista's soldiers and police were running for cover and shouting. The Captain let go of all lines, and we made a hasty retreat out to sea until the rumpus subsided. Later the following day we again tied up alongside the quay but kept a wary eye out for more trouble. That was the one and only time we had to leave in a hurry but it was enough for us.

Around this time there was a surplus of ships and tankers due to the oil crisis and cargoes were difficult to come by. We had just arrived in Jacksonville, Florida with a cargo of molasses and were ordered to tie up, and lay up the ship, as no cargoes could be found. It was a big shock to all of us and most of the crew were discharged and sent home. On board there were only a few of us left. The Captain took leave himself and this left Colby Jackson as the Senior Mate. I too was retained with the Third Engineer and Second cook. The Pumpman made up the full complement of crew on this large ship.

Colby got his wife to come over from Grand Cayman and she stayed onboard with us. As well as being on full salary, we were given substantial daily subsistence allowance, which we hoarded as there was plenty of food onboard, and we also had the cook. Unfortunately, things changed dramatically one night whilst I was the only one onboard. Colby, his wife, and the pump man had gone ashore. The cook and Third Engineer were also ashore boozing. I saw a police car pull up at the gangway, and a cop came up and asked to see the Captain. I explained the situation to him, and said I was the only Officer on board. He then asked me if I would accompany him to the local jailhouse. I asked why, and he said there had been an accident involving one of the crew members, who had been shot in the neck and that a second crew member was being held for manslaughter. I was very nervous because I did not know what to expect, or which of the crew members were involved. At this stage I was not told that the cook and engineer were the men being referred to. When I got to the jail I saw the poor cook in a cell amongst other prisoners. It turned out that the Third Engineer was dead, and the cook was being held in jail for his manslaughter. The inmates looked a tough bunch and were taunting the poor cook, who was devastated and was crying like a baby. I wished I could have taken him back to ship, but nothing could console him. He explained to me what had happened. Both he and the Third Engineer picked up two girls and went back with them. The cook was drunk and the Third Engineer had a few as well, but apparently not as much. One of the girls produced a gun which she said was for protection.

The cook took the gun and started playing with it. The Third Engineer got nervous and tried to take the gun from him, but the cook wanted to hold onto it. There was a struggle and the gun went off. The bullet entered the Third Engineer's throat, and lodged in his neck. He died soon after. I got on to the shipping agents, who hired lawyers, and eventually the Cook was deported to Aruba, from where he came. I never heard about him again. During this time I had the job of getting the Third Engineer's personal belongings together, to ship home to India. I found photos of his wife and child, and felt dreadful imaging how they would feel. The shipping agents had his body shipped back to India. This had really put a damper on our stay in Jacksonville and we were all very low after this.

The Captain and crew eventually returned when we were chartered to collect a cargo of oil and we took off for the West Coast of the USA, via the Panama Canal. This was a repeat experience going through the locks, watching and feeling the ship lifting, and falling, as the locks filled up and emptied. There was great activity as men with very high-pressure hoses continued to cut away at the sides of the cliffs on the Canal. It was fascinating to watch, as mechanical Mules pulled the ship through the locks. The whole thing was a marvel of engineering.

When we arrived in San Pedro California, I was looking forward to seeing this port for the first time. Ashore, I bought myself a suit and cream tie. I suppose, in retrospect, the suit was OK but the tie, oh dear! It was my first taste of shopping in San Pedro, and there were salesmen outside the shops trying their best to get a sale, or a sucker. The reason I bought the suit was because the fellow offered me an extra pair of trousers free. Anyway I found it great fun, and later used to bring home dresses for Molly, which she rarely liked or which never fitted her. I had no idea of what she would or would not like. However, she loved the towels and sheets that I brought home. You could not get the quality at home in those days.

Days later, on the trip along the East Coast, I began to feel unwell and became very jittery at night. If somebody made a sound, even switching on a light, I got a shock that travelled from my stomach to my throat and into my head. It was very unpleasant, and it happened quite often and I began to worry about it, without having any idea what it might be. Then, when we were in port, I was in the loo one morning and felt dizzy and faint. When I went to flush the toilet I noticed my stool was black as soot. Fortunately, we were in port so I could do something about this.

In Long Beach, California, my friend Dr. Chris Walsh from Cobh, and Dr. Mike Riordan from Cork City, were attached to St. Mary's Hospital. I told Chris about it and he immediately had me in hospital for X-rays and tests and these showed I had an active duodenal ulcer. The Irish nuns and Chris were a wonderful help and took good care of me. I was immediately put on antibiotics and a special diet. It was awful. I was banned from taking tea, coffee, or any types of condiments. I had to consume plenty of milk, bananas, soft cheese, and other tasteless bits and pieces. Though the net result was constipation it did start to help improve the symptoms of my ulcer, making me feel better.

During this period we had another bad accident on board the ship. Tanks were being cleaned and made gas free for welding work to start the following day. The gasses build up after oil is transported and discharged. A certificate of clearance was issued and the ship was ready. Next day, 29th May 1957, I was at the hospital with Chris and Mike, when we heard a loud explosion from the San Pedro side of the Harbour. Immediately, there was great activity in the hospital as ambulances took off with sirens screaming. I heard the fire engines going hell for leather. Chris and Mike were summoned to Emergency, and I headed back to the ship, not knowing that the explosion had taken place there. When I got to the docks there were firemen, police and ambulances around the ship which was tied up at berth 241 in the shipyard. Parts of the tank cover were lying on the wharf, and there was a pungent smell of oil. I did not know what to expect, and even after identifying myself, was not allowed to go onboard for quite some time. Fortunately, none of the crew was injured, but one workman named Frank Rawlings, a welder, suffered third degree burns and subsequently died, and four more were seriously injured. It was another tragic accident that left us all badly shaken.

Apparently, there had been a minor fire in this same tank the previous Tuesday, and this had been put out. Now the tanks were certified gas free and work had commenced, with Mr. Rawlings going into the tank, where he lit his welding machine. BOOM! The tank exploded. Somebody had carelessly opened a cofferdam which is a tank that held fuel oil, after the gas free certificate had been issued for the main cargo tanks, and everybody failed to notice this error. The Captain was called to the investigations which followed, but he was exonerated and he continued as our Captain. I never did ask him about it and he never did mention it.

My ulcer was still a problem and Chris asked me if I wanted to go home for treatment by Dr. Denis Wilson, in Cobh, and I reluctantly agreed, because I could not go to sea without medical backup. Chris spoke to Captain Vice, and told him I should be sent home, and the Captain asked me to consider going into the Seaman's Hospital, in San Francisco, which I was reluctant to do. Later I signed off in San Pedro, California, on July 12th, 1957 where I was flown home on sick leave with an active duodenal ulcer. Little did I know what a shock awaited me when I called to see Molly.

1956; Liberty tanker ss Albert G Brown. My 11th ship

1957 Jack and Molly's wedding

1958; ss Adellen (My 12th ship) and SS Albert G. Brown laid up in Texas

1959; Jack Lynch on tug Turmoil in Gibraltar

1959; Deep sea Tug mv Turmoil. My 12th ship

1959; ss mv Kiaora. My 13th ship

1959; Tug Turmoil towing tanker to italy for scrapping

1974; Our family, Ann, Sean, Jane and Susan

Chapter Twenty-five

Finally, after such a traumatic trip I was home. On arrival back to Ireland, landing in Shannon airport, I decided to get breakfast before my journey back to Cork. I was very disappointed with the breakfast I received there. It was cold and the fried eggs looked worn out. They were semi-hard and did not look at all appetising. In fact I did not touch them and was glad to get on the bus for Cork and then the train to Cobh. The ulcer was not helping my mood or comfort so it was with relief that I eventually arrived home and reported to Dr. Wilson, whose instructions were for me to continue with the diet. A few weeks on the diet again and I soon began to feel much better physically but was still quite edgy.

When I flew home from Texas I had a large suitcase which was very overweight for the plane and the shipping agent said he would put it on a ship named Irish Larch which was due in port and would be calling into Cork on her return voyage. I gladly availed of this offer and sure enough the Irish Larch arrived while I was at home and I duly collected my bag. It had never entered my head that the Irish Larch would be involved directly in my life when I used to see her coming into Cork Harbour before I went to sea.

I went to Cork soon after I felt I was improving and called to see Molly. Her twin, Paddy, answered the door and was delighted to see me, and we began chatting. Then the bombshell was delivered. Molly was engaged to be married. This possibility had never occurred to me. Sure, we were leading our own lives but on my terms. I had not contemplated marriage up to now, and it did not cross my mind that Molly might have other ideas. So what do I do now? Molly was out at the cinema with her mother. This they did most days, either going to the Pavilion or the Savoy to see the latest tearjerker. I decided to go for a walk, contemplate my next move, and decide if I should do anything.

My mind made up, I went back to the house that evening. Molly answered the door. We were both awkward with each other at first, and went into the sitting room to talk. I was sure that Molly's mother and Paddy were straining their ears at the door, trying to hear what was going on. Molly and I sat down, and I asked her what she was doing with the ring on her finger. She hesitated, and just said that she was going to get married. I moved closer to her, held her hand, looked at her blue eyes which had tears in them, and asked her who the guy was. She told me his name and I had never heard of him but I knew his brother by reputation as he was a top Irish International soccer player. My only thought was, 'I have to stop this from happening.' Molly is my girl and I intend to marry her.

"Are you sure that is what you want?" I asked. She stayed quiet for a while and tears began to flow. "Molly, darling will you marry me?" I whispered, as I caught her in my arms. She jumped as if I had hit her.

"But I'm engaged, and can't back out now." she said.

"Marry me darling, and give the ring back to him. There is no shame, or crime in this, pet," I said, as I looked her straight in the eyes.

She threw her arms around my neck and said yes. She promised to return the ring and marry me. My heart leaped and I was overcome with joy. Now, we had to announce this to the family. I had to ask for Molly's hand from her Mother.

"Ma, we did it." Molly said when we went into the kitchen.

A strange look came into her Mother's eyes, and Paddy also looked enquiringly, if not embarrassed.

Ma blurted out, "What have you done, child?"

Then it dawned on Molly and me that Molly's choice of words could mean a number of things. What did 'it' mean?

Molly said, "We decided to get married."

"Oh," said Ma, with a sigh of relief, "that's fine then."

Calm reigned again when I now asked for Molly's hand, which Mrs. Kiely immediately agreed to, and gave me a big hug. Paddy was thrilled.

Returning the ring was not pleasant for Molly, and caused hurt to the other guy who wanted to know who I was so that he could 'fix' me. This never happened, even though he found out about me. The family were quite happy with Molly's decision, and so were mine. I brought her down to see them, even though she had previously met most of them.

I remember in the early days of our engagement we went to the Tivoli restaurant, and I asked Molly what she would like. She ordered tea and a bun. I said I was going to get bacon, eggs, and sausages, and asked her if she did not want to change her mind. She stuck to her tea and scone, and I tucked into my breakfast. Later, she told me she envied me eating that breakfast, but she was too shy to order a big meal. I admonished her for this, but it didn't take her too long to stand-up for what she wanted. This little episode made me realise how innocent we were, and really how little we knew about each other.

When we agreed to get married, we decided to go somewhere quiet, to plan our marriage arrangements. We drove to Little Island. Neither of us had ever been on the Island before, or since for that matter. Why we picked Little Island I have not got the foggiest idea, but I was driving and the car seemed to just go there. We found a quiet spot near the water and sat down and talked, hugged, kissed, and made our plans. We decided that my sick leave period did not allow much time for the arrangements so we would have to move fast. First of all we went into Patrick's Street, in the city, and did the rounds of the jewellery shops, looking at engagement rings and wedding rings. Molly selected a diamond cluster, which she wore with dignity and delight. After selecting the ring we then went and had a lovely meal in the Savoy. This time Molly ate with gusto, and was not embarrassed to select what she fancied. She kept looking at the ring saying, "Are you sure it's not too expensive?"

As I had a fair amount of money saved, I assured her that whatever she wanted she could choose. Molly was so excited and showed her new ring to all her friends. It was time to see those scintillating eyes in action. Boy, did they shine and she looked so wonderful, I felt so proud of her.

We decided that September 4th, 1957, was going to be our wedding day. In all fairness, it was Molly and her family who arranged the whole wedding. I was ushered in to meet a Mr. Nathan, a tailor, to be measured and fitted for a suit. I never felt comfortable in that suit, and I didn't wear it much after the wedding. The whole summer was passing by and I still could not come to terms with the fact that I was about to get married. I really had never planned anything and let everything happen before I acted. It was like a dream. I did not seem to have anything to do but wait for the big day in St. Patrick's Church. I carried on as usual when I was in Cobh. The ulcer was slowly responding to treatment, I did some fishing, and went around with my friends with a notable exception; I was not meeting the trains from Cork anymore.

Molls (my pet name for Molly) and I spent a lot of time together in the run up to the wedding especially going to see films. This seemed to affect Moll's Ma, as she and Molls were very close and it curtailed her daily cinema visits, while I was around. During my married life I was to have many disagreements with my mother-in-law.

As the wedding day drew closer I was getting nervy, and the ulcer started to act up. Anthony was to be my best man, and my parents, sisters, aunts, uncles, and friends were all making plans. I had no idea what to do, except let time go by, and on the fateful morning, got washed, shaved, dressed, and watched in a dream as my family buzzed around me.

Anthony had arrived from Portsmouth, and had travelled part of the way by motorbike. During the trip he fell asleep on the bike and crashed into somebody's garden in Wales. He got up, dusted himself off, left the motorbike in the front garden, and got a bus to Fishguard where he boarded the Innisfallen for Cork, none the worse for his experience, except for a bruised knee and hurt pride. He arrived on time, and did not relate this episode to me until sometime later. He never went back for the bike!

Molly's sister Doreen was to be Matron of Honour and her brother Donald was to give the bride away. Early on the morning of September 4th I got into Joe Twomey's taxi with Anthony and my parents, and we headed for Cork. I was still in a daze and cannot remember the trip, and even when we got out of the taxi it was unreal to me. At this time the duodenal ulcer gradually began to tell on me, and I noticed myself getting more and more nervous and tense. We called to Molly's house and were ushered into the front room, as Molly was upstairs with Doreen and Anne who were helping her get dressed. Mrs. Kiely asked me where my buttonhole was and then it dawned on me that all of us should have buttonholes and nobody did. Panic! Mrs. Kiely had only enough for her family's side, so she had to take a flower here and there from the others to make up the extra buttonholes for me, Anthony, Mammy and Daddy. All fixed up, we headed across the road to St. Patrick's Church, which was directly across from the house, and took up our seats.

After what appeared to be an eternity of looking around, and fidgeting, the organ started to play, 'Here comes the Bride.' This is when I felt it was not a dream, but reality. I glanced back and saw Molly's eldest brother Donald leading her up the aisle. I moved from my seat to the centre of the aisle. I could not see her face, because the veil covered it. When she drew up alongside me I put my hand out, and gently squeezed her hand and smiled. She smiled back, and it seemed as if this was the most natural thing to be happening to us. My fears and doubts were gone, and I felt at ease. Even the ulcer behaved. She looked enchanting, radiant, and beautiful. The ceremony commenced, and after we had taken our vows and signed the register we went outside for more photos to be taken. We then retired to the Metropole Hotel for our wedding breakfast.

Unfortunately for Molly and me, the Heffernan Travel agents had not brought our travel tickets and we wondered if we were going to miss our planned honeymoon; Konigswinter, on the Rhine in Germany. At the Metropole Hotel an agent arrived as we sat down for breakfast. This left little time for us to sit down at the table and enjoy our meal. It only gave us 1 hour to eat, get ready and go to the station for our train. We ate nothing due to excitement and the poor arrangements. When the speeches were quickly finished we headed for Cork Railway station, to catch the train for Dublin and make our ongoing connections. We were bitterly disappointed with Heffernan Travel agency for making such poor arrangements, but nothing could be done at this stage. We had been chasing them for the tickets and were in fact fearful they would not arrive at all in time for us on the day. It was ironic and heartbreaking that on our wedding day the tickets arrived and Molly and I had to bid farewell to our guests far too early, we were later told they had a great time and thoroughly enjoyed themselves with lots of songs and craic.

Molly and I got to the railway station in time, and here began the most tiring and frustrating part of what was to be our most enjoyable experience. I was the first to start this off by going to the Kiosk for something for us to eat, as we had little food since the evening before. We had to leave without our own wedding breakfast. To my utter shame and thoughtlessness, I bought a sixpence bag of Bon-bons. I returned to Molly in the carriage and sat directly opposite her at a window seat. There was nobody else in the carriage at the time. I took out the bag of Bon-bons and asked Molly if she would like one. I got a fiercely disbelieving look and a frosty reception; as only Molly could give, and a blunt, "No."

She told me later she was starving and was waiting for me to come back with a box of chocolates, or something nice, but a white paper bag of Bon-bons. It beggars belief. To this day I smile when I think of how naïve I was then, and wish I could have turned the clock back. We said very little on the way to Dublin, and only when we had a meal later that day did things return to normal, and after I apologised, and we hugged and kissed, we were fine again. However, she never let me forget her first meal on her wedding day!

After waiting hours for a flight, we now had to fly to Brussels in Belgium and arrived there totally exhausted. Worse was to come. There had been very bad weather over Brussels. All helicopters were grounded for the previous twenty-four hours. This was a surprise bit of news to us as nobody had told us we were to fly by helicopter to Cologne. Neither of us was ever in a helicopter before and we were both apprehensive at the thought of it. Molly was more frightened than I was but we had to face up to it. When we eventually went to the helicopter pad we were told we were the first flight up for twenty-four hours. This did not instil much confidence as there was still a wind blowing as we got aboard the Sabena chopper. As the time for takeoff approached, I looked around at this small craft. There were seats for three people facing another three seats, and there was only one other seat occupied by an elderly man.

The chopper engine started up, the rotors began to whirl, and the chopper started to shudder as it began to lift off the pad, swaying and jolting. I did not like the sensation one bit, and Molly kept coming closer to me whispering how afraid she was. I tried to console her and gradually the chopper levelled off, and after some time we thankfully arrived safely in Cologne. All this time I had my ulcer for company!

Our next problem was how to find our way to Konigswinter. As we could not speak German we were having difficulty making ourselves understood. As usual Molly left me with the problem of trying to explain where we wanted to go.

"You ask," was the usual request, or was it a command?

Anyway, we found our way to the Cologne railway station after much difficulty. Heffernan Travel Agency had told us we'd be met there by their representative. No such person was there, or if there was, they did not make themselves known to us. As we did not have a phone number or address to contact we now had to figure it out ourselves and make the best of it. Molly, by this time was getting very upset with me because I was not resolving the situation, and she was very tired. Neither of us had a wink of sleep for over thirty hours, and we were just about ready to snap each other's heads off. Some start to our honeymoon!

I eventually figured out where to get the train to Konigswinter and we dragged our bags into the carriage. The journey took about thirty-five minutes, and we arrived too tired to notice the Rhine or anywhere else. All we wanted to do was to get our heads down and grab some sleep.

At the Hotel Loreley, in Konigswinter on the banks of the Rhine, we booked into room No. 50 on the second floor. There were no lifts and nobody was available to bring our suitcases upstairs. All the steps were white marble with no carpets and this was a killer on the feet. After we unpacked our cases we fell on the bed and went straight to sleep. It was hours later when we woke up. This sleep did not help too much because of wedding night nerves, tiredness and hunger. It was all too much for Molly and she got up, looking very upset and said; "I want to go home to Mam."I too was very tired and fed up and said, "OK, go home, and stay with your Mam."

"Where's my passport?" Molly asked.

"Where you put it," I replied and turned on my side on the bed, still tired. Molly went out the door with her passport. With the light outside, and the gap under the door, I could see her standing just outside the door. I thought to myself, 'Where does she think she's going without tickets, baggage, or knowledge of the place.' After about ten minutes she appeared back inside the door and I pretended to be asleep. She came over, stood at the bedside, and looked down at me. She whispered, "Jack, I'm sorry, I'm tired and hungry."

I felt a deep sense of love for her at that moment and stood up, and put my arms around her, and told her I loved her very, very much. She immediately relaxed and both of us, feeling much better decided to go out to get something to eat.

The hotel was more or less closed for the night so we had to go to the shops. We ended up with a bottle of Mosel wine, crusty rolls, butter and cheese. We rushed back to the hotel and I started drinking wine for the first time in my life. This was not on the menu for my ulcer but what the hell! See what a woman can do to you! We grew to like Mosel wine, but it was the first time we had seen blue cheese and it gave us a jolt. When we opened it and smelled it, we thought it was gone off so we threw it out.

Each morning in the hotel we usually had a continental type breakfast, but at lunch we got fed up with the same dessert every day- stewed apple and custard. It got to the stage that we refused to take it and the poor waiter got very upset. He couldn't understand why we did not like stewed apple and custard and we couldn't explain to him that we did not travel to Germany to get what we already were used to getting in Ireland.

We thoroughly enjoyed exploring Konigswinter and the surrounding cities for the two weeks of our honeymoon. We took lots of photos in Konigswinter, which was small town. It was very close to Drachenfels, which was very high up overlooking the Rhine, and from here we got great views. There was a miniature zoo there, and I nearly lost an eye while visiting this zoo. In a cage there were some monkeys, and I did not know how dangerous it was to go too close to the bars of the cage. As I approached, to get a closer look at the inmates, one jumped down close to me, and with a scream he pushed his paw through the bars and cut my nose just close to the eye. There was small cut and I was worried I might get some infection. As it happened, there was a small stream close by and I bathed my wound there to my great relief. Molly was concerned, as she thought he had injured my eye. Apparently, it was a male monkey protecting his female friends. That's what you get if you poke your nose where it's not wanted. I was much more careful after that escapade.

We visited the Hotel Petersberg, which was close by and overlooking Konigswinter. This is a magnificent hotel, and very posh. The grounds were beautifully manicured, and we spent hours walking around, looking at the views of the Rhine. Apparently, Adolf Hitler and his top brass used to meet here quite often during the war. When we went in for a look, Molly said she would like a cup of coffee. We ordered two coffees and got a shock when we were charged fifty pence (it was usually twelve pence back at home) No more high-flying for us in the Hotel Petersberg. The experience was worth it for Molly but I thought it was extravagant.

Bonn was a beautiful city and we found it was easy to get there. We went there a number of times and took the cruiser-boat Siebengebirge up the Rhine, from Konigswinter to Bonn. It was terrific to see all the different cruise boats, barges and ships moving up and down the river. Other times, for a change, we went by the train. We also went into Cologne by train and visited its cathedral. This was a beautiful cathedral, but had been damaged during the war. The 4711 Eau de Cologne shop fascinated Molly, and we strolled around and returned there on a number of occasions.

Two other towns we went to see were Bad Godesberg and Bad Honnef. In Bad Godesberg, the day we were there was pre election day, and the cars were decked out in flowers and had loud hailers. Something similar to Ireland, in the old days!

In all these cities and towns, we found the people very friendly and helpful. We visited lots of cafes, and drank lots of coffee and Mosel wine, and generally had a wonderful honeymoon. We got to know each other's little failings, and began to forgive more easily, instead of jumping off the deep end when we felt we were being wrongly accused of some trivial absurd failing.

The time came to return back to Cologne airport, to board our helicopter again for the return to Brussels. This time we were much happier and not concerned as the weather was fine and the journey went without a hitch. We boarded our aircraft and took to the sky. The next stop was London. From Heathrow Airport we went to central London, where we booked into the Regent Palace Hotel. We stayed there two days and went around London sightseeing. We went to 'Madame Two Sods,' as I called it. Molly kept correcting me about the pronunciation. Eventually, it was time to head for home. We returned to Cork, via Paddington and Fishguard. In Fishguard we boarded the Innisfallen, to cross the Irish Sea to Cork. How lovely it was to come in past Roche's Point lighthouse and then the Spit lighthouse. I could see my family waving from the house as we passed up the river Lee to Cork. It was a great thrill to see most of the people of Harbour Row waving to the boat. The people did this every voyage, as the Innisfallen was a focal point for the people. In Cork, Ma and Donald met us as we disembarked from the boat. The family home was only five minutes' walk from the berth where the Innisfallen docked.

As I was still on leave and we and did not intend to buy or rent a place until after I retired from sea, got a shore job and we got settled, Molly suggested that she would stay with her family. When we got to the house Molly had to give a detailed run down on our trip and honeymoon, but of course was very discreet about our private lives. Ma would have liked to hear more of this side, I could tell by the probing questions she asked. Later, she referred to me as a Stallion and her other son-in-law as an Aberdeen Angus. I could feel it was going to be embarrassing for Molly and me, on our first night in bed together in the house. This was the beginning of a strained relationship between my mother-in-law and me, which lasted for quite a number of years.

It wasn't long before I was required to rejoin my ship, and the day arrived too soon for us. I had decided I would stay at sea a little longer before I was ready to finally 'swallow the hook' and retire to land life. This was a difficult decision as I wanted to save more money and it was with a heavy heart I said my goodbyes to Molly and headed off to join my ship in Rotterdam, Holland. It was October 1957, six weeks after I was married and I was to leave my new bride for who knew how long.

Chapter Twenty-six

When I rejoined the Albert G. Brown, unhappily so, as I found leaving Molly much harder than I expected, but had no idea that we would be reunited sooner than I could have imagined.

This time on the AGB I had a different Captain. He was the Captain of another of the Company's ships, Kiora, which was laid up in Houston, Texas. He was acting as relief for Captain Vice. I had not previously met him but I liked him immediately. He was a Dutch/American man whose passion was for music, especially classical, which he played at every opportunity. He was excellent at Mathematics and asked me if I would like to take up higher mathematics, which he would teach me. I declined, and to this day regret I did not take him up on his offer. The trip was easy enough and mercifully quick. We did another trip to Cuba and Jacksonville, Florida and returned to Rotterdam around December.

Molly and I, missing each other a lot, made plans for her to come and meet me in Rotterdam when the ship docked. I was so proud of her and organising ability in sorting out the travel arrangements and her bravery on coming over alone to meet me. It was her first trip abroad alone and it took some deep courage in those days when people did not have the comforts they have today with modern travel facilities. When I wrote to her outlining what she would need to do to meet the ship in Rotterdam, she followed my instructions faithfully, and for someone who had never travelled much on her own, I did not realise what I expected of her. She had to arrange her tickets and times for the trip, by boat and plane, from Cork to Rotterdam. She had to change money in the bank, make all the bookings and organise all the paperwork. It took her a day of travelling to get to me.

She flew to London, then got the train to Harwich and arrived via the Hook of Holland. There she boarded another train to Rotterdam. My poor darling was all alone, could not speak Dutch, but coped magnificently. She did have a scare when she was trying to contact the shipping agent in Rotterdam, found the office was closed, and could not contact anyone. She must have looked lost, because a small elderly priest asked her, in English, what her problem was. She explained to him about meeting me, and that her contact was not available. He asked her to follow him, and that he had a sister living close by, who would help her. Molly without fear followed him, and he took her suitcase and carried it for her. After going through a number of small back streets, Molly began to fear that this man might not be a priest, and that he had ulterior motives for helping her. She tried to think of a way out when they arrived at a small house and an elderly lady came to the door. She listened to the priest and smiled, took Molly by the hand, and put her up for the night. In the morning she phoned the Rotterdam office and an agent came to pick Molly up, and told her my ship was not due until late that night. Those lovely Dutch people who helped Molly will always stay in my memory. Molly later told me that the ship's agent tried to take advantage of her but she defended herself with dignity, telling him she would tell me. Just as well it was some time later, before I heard about this!

When our ship arrived at the berth in Rotterdam it was about ten p.m and very cold. I stood on the bridge as we came alongside the quay. The ship was not carrying any cargo so it was riding high, and the wharf was way below us as we tied up. As I looked down from the bridge of the ship I saw this forlorn slip of a girl standing alone, with her suitcase, and looking up at me as we drew alongside. She looked cold and my heart went out to her. I called her name and she waved back. What a brave girl I had married. I was so proud of her and more so when I ran down the gangway I hugged and kissed her, before bringing her to my cabin. When all had calmed down and the other more experienced wives of officers had arrived, we all set to introducing ourselves to each other. Captain Catlender was very good to Molly and made her feel at home. She too, liked him and I was pleased that everything looked to be going well.

We did not have much time in Rotterdam before we set off for our next port, Hamburg. Molly liked to spend time on the Bridge, looking out, enjoying the sea, and taking in all the activity onboard. I was on duty in the Radio room for much of the trip but luckily it did not take long, and it was lovely having Molls aboard. I was quite friendly with a Maltese engineer who was familiar with the night-life in Hamburg so one night we took him up on his offer to take us ashore to a nightclub he recommended. I had never been to one before, and Molly certainly had not even known this life existed. To our surprise, when we arrived, it turned out to be less of a nightclub and more of a nude woman show. Poor Molly did not know where to put her face. Girls came out on stage in their birthday suits and danced around the floor. What really got Molly was when one girl, who had a feather duster, kept dusting another girl 'downstairs'. Molly buried her head in my back in embarrassment and eventually we had to leave because Molly could not take any more. It sure was an experience for both of us.

We went ashore a couple of times more on that stopover and we enjoyed each other's company before the time came for us to say our goodbyes as all the wives were going home. Fortunately, two of them were going by train to the Hook of Holland, so Molly had company for some of her return journey. From there Molly went on to London, where she stayed for a while with her sister, Anne. Being Sunday when she arrived there they decided to go to Mass. Instead of going to Westminster Cathedral, a first time for Anne too, they ended up in Westminster Abbey. It took them some time to realise their mistake and they had a good laugh about it. From London, Molly travelled home and continued to stay with her mother while I was at sea. My ship carried on and we sailed back to Houston in Texas, where the ship was again laid up, due to lack of cargo. On the 24th January, 1958, I flew home again, on full pay, to await further developments. It was great to be with Molly again, and we made most of our time together. I never again sailed on the Albert G. Brown, and have only now found out what happened to her.

July 2005; I have just located the following on the internet and it brings back good memories, but still leaves me sad at the ending of the AGB:

History of Albert G. Brown.

1943. September 3rd Keel was laid.
1943. October 18th ship launched.
1943. November, Built by Delta Shipbuilding Corporation, New Orleans, Louisiana.
1943. Operated by American Trading and Production Corporation, Baltimore, Maryland.
1948. Owned by; Bernuth Lembcke Company, New York.
1954. Owned by; Torres Shipping Company, Liberian Flag (Bernuth Lembcke Co. New York).
1960. Scrapped Hirao, Japan.

Sad! I remember somebody once said that ships that are scrapped could end up as razor blades. I wonder...

For five months I drew full pay whilst remaining at home. It was fantastic and felt wonderful to have so much time to enjoy with my new wife. During this time, Molly and I stayed in Cork for some of the time. We also travelled around, went to London, and visited various other places. All in all we had a great time. We spent some time at home in Harbour Row with mammy, daddy and the rest of the family, which I really enjoyed.

Then I got the call to go to South Shields, in England, to help commission another tanker that the Company had purchased and life was to keep moving on again. There were more sad goodbyes and promises to write as I packed my bags and got ready to leave for another unknown duration.

Chapter Twenty-seven

As I headed off to my eleventh deep sea voyage, this time on board the s.s Adellen to USA, Aruba, Curacao and Venezuela, It was June 13th 1958 when I put pen to paper and signed on and I learned I would be joining the crew from the Albert G. Brown. It was good to meet old friends again, including Captain Vice, and when we were all signed on in South Shields and sailed to Newport, Monmouthshire, in Wales, where the ship was to be totally overhauled. This would be a long stay again in port. The ship had previously been British owned and all the fittings and supply lines were also British. However, it was now owned by the Torres Shipping Company (Bernuth Lembcke Co. New York.) The Chief Engineer took great delight in replacing all British fittings with American equipment.

This turned out to be a momentous time in my life as it was around this time that Molly became pregnant. I remember when I phoned home she told me that she felt she was going to have a baby, and waited for my reaction. I don't know what she expected, but I was overjoyed. Within a few days I came home on the Innisfallen for a week and the second I saw her I picked her up and hugged her with delight. Then, I thought I might hurt the baby so I calmed down and placed her gingerly back on the ground! Both our families were delighted and we had a big celebration. Molly seemed to bloom more and more as the days went by. After a week or so I had to return to the ship for completion of commissioning.

Back on board it was all hands on deck. As well as having the engines and operating systems inspected, torn apart, and then reinstalled, the Chief Engineer continued to replace all electrical wires and fittings with American replacements. Finally, the day came to take our ship out of dock and go for trials. We moved slowly up the river, and as speed was increased the engines froze, and shut down. We needed assistance. There was no way we would be able to get underway under our own steam. The Captain came to me and asked me to get an urgent message off to the agents, asking for tug assistance. As we were under pilot control, and had not left port, I was technically not on duty. Nevertheless, I switched on the transmitter, and started to send the message to Seaforth radio, which was the closest medium wave station. Alas, nothing was happening. The aerial was grounded somewhere, and since we were not leaving port I had not felt it necessary to check before we left the dock. After a quick look at the aerial, I realised that the only place something was wrong was within the steel box, which housed the down lead, via the insulator to the Radio Room. To access the inside meant that about twenty-five steel nuts would have to be removed, and repairs would have to be carried out to the aerial. This would take too long, so I had only one option and hoped that it would work. I advised the Captain of the problem and he was up the wall. With his consent, I began to put into action my plan. I switched to short-wave radio, on 8 m/cs and tried for Portishead Radio, GKN, which was only a few miles away. I hoped that the signal would get through using the short section of aerial between the transmitter and the isolated section to the main aerial. Out went my call to Portishead, and I could not believe it when they came straight back to me and confirmed I had got my message off OK. Soon after, a tug arrived and took us in tow back to the dock. Further repairs were carried out, and as I suspected the aerial problem was within the isolating box. It turned out that welding had taken place on the box, and the heat had melted the solder at the internal connection, causing the aerial wire to fall off, and become grounded. When this was repaired everything was fine. I was pleased I had managed to sort the problem but not as pleased as the Captain!

After the remaining repairs were completed further trials proved successful and we headed out to sea. We were bound for Vita, Cuba, where we loaded Molasses, and took it to Jacksonville, Florida. We had another short trip to Curacao, and Venezuela and back to Jacksonville before the ship was again laid up and I signed off on December 10th, 1958. This time again, I got home on full pay and standby. It was a lucky and exciting time as I awaited a very special arrival.

I had been in frequent contact with Molly all this time, either by ship's telegram or letters so though I had been hearing all the exciting news and developments concerning the pregnancy, coming home and seeing it for myself was very special. We both looked forward to seeing each other and when I did arrive home it was terrific. I could not believe my eyes when I saw Molly this time. She had been working around the house, getting things ready for my return and did not expect me home so early so was not dolled up as she liked to be. To me, she looked radiant. The changes, though, were astonishing. Her once slim waist now showed the change which occurred over the past eight months. There nestled inside, was our first baby, and the thrill of feeling the movement of this baby was wonderful. Was that a bum or a leg which just moved? It was a great guessing game! We spent hours going for long leisurely walks and talking all about the exciting future ahead of us. We wondered about the baby's sex and planned the coming years ahead with excitement.

Molly's sister, Doreen, was also pregnant at the same time and I was pleased that Molly had someone else to discuss all the intricacies of her new found condition with. It was January 1959 when Molly started getting her contractions. Her Ma kept monitoring the times and frequencies of them whilst I kept asking if it was time to get a taxi to the hospital as I was eager to leave and get to the relative safety of the medical establishment. I was getting quite nervous, as the contractions became more and more frequent. We had realised that labour had started earlier that night when we were in bed and Molly had given a yelp as all of a sudden the waters burst. Ma arrived into the room, while I fussed about wondering what to do. Ma took over completely. She told Molly that it would loosen her up for the birth if she walked to the Nursing Home which was reasonably close by. It was about two a.m on a cold January morning, when we started out from the house and walked slowly up Summerhill. Molly was in between her Ma and me, and was very brave chatting and smiling all the while. I, on the other hand, was a bundle of nerves. When we reached the Nursing home we were pleased to discover that Molly had a lovely room to herself. The lady in charge said there was no point in us waiting all night as she felt Molly was not quite ready to deliver the baby so she suggested that we go home.

Ma and I walked back to the house, after saying our goodbyes to Molly, and promised to come back up in the morning. We had a cup of tea and a cigarette, and then went to our beds. Eventually, I fell asleep. I awoke early in the morning with a mixture of excitement and trepidation wondering what the news would be. As we did not have a phone in the house we had no idea what was going on in the nursing home. So, after waiting for a while more, and a couple of cups of tea later, we headed back up the nursing home, with me praying that it was all over, to find out to our delight that Molly had given birth to our first baby.

A smiling maternity nurse met us, and told us that Molly had a beautiful baby girl, who was born in the early hours of the morning, soon after we had left the home. We hurried into see them both. Molly was sitting up in bed in the room, looking radiant, and smiling from ear to ear, I immediately went to her as the new Gran went to the cot where the baby was asleep. I kissed Molly, hugged her, and told her how proud and relieved I was to be a father for the first time. In the cot beside her was a lovely, rosy-cheeked baby, who was trying to get some sleep, despite all the excitement around her. I picked her up gently with awe. She was a little gem, and I hugged and kissed her, and welcomed her into the family, and into the world. I felt so proud. Molly was laughing at my antics, as I looked at the tiny fingers and tiny toes, marvelling at this wonderful gift from God. I put the baby back into the cot and held Molly's hand. She told me that she kept counting the baby's fingers and toes to make sure all were where they should be.

The room was bright, and there was a lovely fire in the grate. Molly looked so relaxed, and said the few hours of sleep she got after the birth was the nicest sleep she had ever had. She was to get those sleeps on three more occasions! However, there were many other nights later, during the rearing of our children, that were far from nights of pleasant sleep as we were kept up feeding them, changing nappies, and helping them with teething and other illnesses. It was worth it all.

When Molly finally came home with our baby, there was great excitement. Donald and Paddy were over the moon. Donald was to become her chief protector when I was not around and he loved playing with her. We took our little baby to Cobh to visit my family and there was a lot of commotion when we arrived at home. Daddy and Mammy were thrilled, as were my sisters. Anthony sent his best wishes from the USA. Daddy used to drive me up the wall by calling the baby "Lovely Girl," over and over. It was always in a singsong manner. He loved her, and it was his way of expressing this love for her. Another admirer that she had was a dog called Sparky who loved to sit and guard the pram. Nobody, except family, was allowed near the baby, her watch dog. Sparky, growled and barked to frighten people away from our precious new treasure.

We now had to think of a name for our baby and it did not take too long to decide on Ann, in honour of Molly's mother. Marie was added as a second name, though there was no relative who bore this name, it was just that Molly liked it to go with Ann. So our first baby was to be called Ann Marie Lynch.

We bought a lovely high pram, and Molly loved to dress Ann up and keep her in spotlessly clean baby clothes and walk her through Cork city, meeting her friends, who stopped to admire baby Ann. We kept taking photos on a regular basis. Soon enough, as is the Irish way, the day for the christening arrived, and of course relatives and friends congregated at St. Patrick's Church, where Molly and I had been married, to welcome our daughter Ann who looked lovely in the christening gown, which was a family heirloom, and had been used in the christenings of all her side of the family, including Molly herself. Ann behaved very well, and did not cry or make noises, even when the cold water was poured on her head. Ann's Godparents were her Uncle Donald and Aunt Anne Chapman (née Kiely). Both families attended this special occasion and we all retired to Molly's house for tea and cakes after. It was a wonderful day and a special time, however, it was after this lovely occasion that things began to change in the house.

Everything was going well enough except I was finding it difficult to accept Ma's ways, and I felt she tried to take over a lot. She was often issuing orders to me and, since she had three sons who had to obey her every whim, she felt I too should do the same. However, I too had a strong will and was not used to being humiliated and bullied. Anyway, the long and the short of it was that one day, whilst in the kitchen having tea, I asked Molly if there was a clean shirt for me to change into for the following day. Ma interrupted and started to give out to me.

I got up from the table and left the kitchen. There was dead silence from all remaining and I felt sorry for Molly being stuck in between us but I was not going to be subjected to this kind of treatment. I called for Molly then and told her I was leaving the house, and would never step foot inside the door again, not until I got a genuine apology, and even then that I did not expect to sleep there again.

I asked Molly if she was willing to come and live in Cobh with me, until we got settled in a home of our own. Poor Molly was torn between her mother and me but said she would come with me. I think Ma believed that I would be back in a day or two, but I told Molly to tell her that I would have a removal van at the house the following day to collect our things. The shock on Ma's face, and on Donald's, when we moved out was something I remember to this day. Things were easier living with my family in Cobh. We settled in quickly and enjoyed being a family. During this period baby Ann was able to sit up, and Molly used to sit her on a blanket outside the house, and Sparky would sit close to her so she could reach out and touch him. He was so patient towards her and he would occasionally lick her little face and she would gurgle and laugh with delight. A beautiful sound!

Molly used to go up with Ann to see her family a fair bit whilst we were in Cobh but I never went near the house until a long time later, when Molly asked me to meet her mother. She said her mother wanted to make up and I reluctantly agreed to also for Moll's sake, but I never again intended to spend a night in the house. Ma was true to her word, we both made up and as we later moved on and got our own home she used to stay with us on many occasions.

During my period of leave when Ann was a baby and we were living in Cobh I loved to go down to the Deepwater Quay to watch whatever ships were tied up there and one day I watched as the tug m.v. Turmoil pulled in alongside. She had been stationed in Cobh for many years and was a very popular sight as she had been involved in the famous, but futile, attempt to save the American ship, the Flying Enterprise. The Turmoil was manned by mainly Cobhmen, and had two Radio Officers on board whom I knew - Tony Byrne and Sean Leahy. Sean asked me if I would standby for him, as he wanted the weekend off to go to Cork with his wife. There had to be two Radio Officers on board at all times so I thought 'Why not?' It was only for two days at most and Bernuth Lembcke would give me at least a week's notice when they needed me. I figured it would be a simple enough job. I went home and told Molly, who was not too worried either. All discussed, I decided to sign on, and met Ger Kelleher, the Shipping agent, who was a good friend of mine as I signed the dotted line. I had no inkling of what lay ahead for both Molly and me and the anguish we would suffer as I signed on this ship. I expected to just sit aboard doing a watch keeping spell for the weekend but providence was about to intervene.

Chapter Twenty-eight

After I signed on the m.v. Turmoil, on the 6th May, 1959, I familiarised myself with the ship's call-sign; GMWK and I took up Radio watch on the distress frequency, which was maintained twenty four hours per day, even when the ship was in port. I was relaxed and at ease, expecting this to be a very simple and easy couple of days but after about five hours we got orders to proceed into the Atlantic, where a ship was in distress. Despite my assurances to Molly that this I would return home in a two days this turned out to be my thirteenth deep sea voyage.

The Turmoil, which had been built 14 years previous, was chartered out to Overseas Towage and Salvage Company. She had been built by Henry Robb Leith of Scotland for the Admiralty and her gross tonnage was 1136.13. Immediately after the distress call was received the ship's hooter sounded off, calling the crew from home and only giving them fifteen minutes to board as we prepared to cast off. If a call back was at night time a message was flashed on the screens of the local cinemas also, to ensure all possible avenues of communication were covered. Word soon spread around the town that Turmoil was again going to sea on a rescue mission.

I got word to Molly quickly, who was looking out of the window from Harbour Row and I watched her waving a white handkerchief as we steamed by, rounded the Spit lighthouse, and headed out to Roche's Point. Baby Ann was with her, and Molly later told me that Fats Waller was singing Blueberry Hill' as we passed her by. I had tears in my eyes, but felt we would not be away for more that a few days, but had the added worry that Bernuth Lembcke might call me for work while I was at sea.

As we approached the distress area we had a message saying another tug was already there and we were not required. I was delighted and thought. 'Great! Now we can go back again, and in a few hours I'll see my family.' How wrong I was! A further message told us to proceed to Antwerp in Belgium, and pick up a large French tanker, named, s.s. Bourgogne, and tow it for scrapping to La Spezia, Italy. We headed into the English Channel and the North Sea towards Belgium to fulfil our contract and I was very annoyed because I did not know how long I would be away from my precious girls.

    We collected the tanker and began towing it in calm sea through the Channel but when we got towards the Bay of Biscay the weather took a turn for the worst and the waves began to get bigger. Fortunately, it was not a full gale, but it was frightening. I had no experience of this kind of work before and though I found that the tug was powerful, despite its small size, when I saw the towing hauser fully taut, with this large hunk of ship behind, I wondered what would happen if the hauser snapped. I tried to visualise the wire snapping, and then hearing a cracking noise as it came lashing towards us, cutting everything in its way. The most frightening aspect was when we were down in the trough of a wave, and the ship behind loomed high on the crest of the following wave. Then, as the tug climbed the next wave, the looming ship started to descend down the last wave. I felt she would come crashing down on our tug and it was frightening. Thankfully, it never happened, and the hauser kept automatically adjusting to the strains, and stresses of towing. We made good time and soon passed Gibraltar and headed into the Mediterranean which was calm and sunny.

    When we arrived in La Spezia, Italy, we unhooked from the tanker and restocked our larders. Our stay was short but I got ashore for a short walk around. It was a beautiful place and on this day the sun shone brightly and it was quite warm. There were a lot of naval personnel around as this was a large naval port. The place looked like a tropical holiday resort, and though I was hoping for more time there I was also keen to head back towards Cobh, which looked to be happening very soon, or so I thought.

    The next day another message arrived telling us to proceed to Milford Haven in Lancashire for further orders. Although I was disappointed not to be heading straight back home I was relieved that at least I could get off in the UK, as I did not want to push my luck and I hoped Sean Leahy would meet us there and replace me allowing me to return home. In Milford Haven the agent arrived, gave us our mail and issued us with a new set of orders. The tug was to go across the Atlantic and proceed to Baltimore, Maryland to assist other tugs tow two obsolete American aircraft carriers to Japan. Without a second thought I signed off immediately. I was hoping the tug would return to station in Cobh but no such luck. Most of the lads signed off here in Milford Haven too and we all went home together. There was no way was I going on this long voyage to the United States with my new baby Annl waiting at home.

A number of the Cobh lads accompanied me home and we had great craic on the journey back. None of them fancied going on the trip either and it was great to have the company on the trip back to Cork. It was wonderful to be back in Cobh with my family again, and it was another month of bliss before I was again called to go and rejoin my next ship. Ann seemed to have grown even more over these weeks, and Molly looked radiant. I felt I never wanted to leave them again but my sea life had not yet ended and I had to return to fulfil my destiny.

Note on the history of the Turmoil.

I've since researched the ship to find out its fate and have learned that the m.v. Turmoil earned her place in salvage history during the Flying Enterprise saga, and was stationed in Cobh, until December 1959. In 1965 she was sold and had her name changed to Nisos Kerkyra; which is a port on the coast of Corfu, Greece. In 1971 her name was again changed to Matsas, by the firm Loucas G., the largest Salvage and Towage Company in Greece. She was painted white and emblazoned with the Maltese cross on the funnel and she was finally broken up at Perama, Greece in January 1986.

Chapter Twenty-nine

My destiny was to take my back to the United States again, this time on board the tanker m.v. Kiora on what was to be my 13th, and last, deep sea voyage via Rotterdam, the Panama Canal and Seattle. The ship, which was also owned by Bernuth Lembcke Company, New York, was a WWII T2 Tanker and was registered in Monrovia, Liberia. I signed on in Killingholme, Scotland on July 18th, 1959 yet although I signed on in Scotland I actually joined the ship in Dover, England where I had been told to report to join the Kiora, as it passed up the English Channel. I waited for three days for the ship to arrive and thoroughly enjoyed myself there. The shipping agents were in constant touch with me, and eventually advised me that the ship was due to pass Dover the following day. The Kiora was on its way to Rotterdam, Holland and had another Radio Officer on board who I would be replacing. I rejoined it from the Pilot launch, which took me from the port, out in the Channel, to the ship. As there was another Radio Officer onboard, I was therefore classed as a passenger, or supernumery, and did not take over duties until the ship arrived back in Killingholme, Scotland. Once we hit Scotland the other Radio Officer signed off and left the ship. Apparently, he had only been employed for this trip. I commenced work at this point.

    Captain Catlender, whom I had sailed with earlier on the Albert G. Brown, was in command and was his usual hearty self. It was great to work with him again and we carried on as if we had never been apart. It was strange because Captain Vice never again came into my life, nor did Colby Jackson. I still think of them both and often wonder what happened to them.

    Our trip from here took us across the Atlantic, to the Panama Canal, and up the west coast of the USA. The ship headed for Seattle, Washington and arrived there on August 22nd, 1959. Again, we only had a short time there before we had to leave. I spent some time fishing for salmon but with no luck, a far cry from my youth when I never missed a bite. My mind was constantly on my family and I had already decided that this trip would be my last. I posted my usual letter to Molly when we arrived and was pleased that there was one there for me as well. Everything at home appeared to be great and I was looking forward to getting back and settling down. On the way back it was all plain sailing, as the saying goes, and when I returned to Grangemouth and signed off on November 10th 1959, my sea going days had finally come to an end. I travelled light on the way home, with a burning longing to be with the people I loved.

I returned to my family, prepared to enjoy myself as I practiced becoming a landlubber after 10 years on the high seas. All in all, I had a great time at sea. I felt that all the leave I had with Bernuth Lembcke had prepared me settle down, and get over the longing to remain at sea. Some other fellows were not so lucky and could not stay ashore for any long length of time.

I was ready to embark on the next phase of my life; one that was to hold many moments of love, affection, grief, fun and joy, one that was to create many new and wonderful memories for me to treasure beyond the sea.

Chapter Thirty

Molly and Ann had moved back to Cork while I was at sea and I did not want to upset Molly by asking her to come to Cobh with me when I packed up my life at the sea. I therefore decided to swallow my pride and on assurances of Ma that there would be no more interference I stayed with her, and lived in Cork. While there, we had a very upsetting episode in our lives. Ann was still a little baby when Molly became pregnant, and suffered a miscarriage. Her Ma, fortunately, took over as usual, and I believe the baby would have been about five weeks at the time of the miscarriage. We were distraught but Molly and I got over the shock, and we settled down to our normal life. However, there was no work at home that suited me so we decided to embark on a journey that would eventually take us to London, Dundalk, and Dublin.

The three of us got the Innisfallen to England and settled at number 28, Hanger Lane in Acton, London. We rented a one room flat. The landlady was at first reluctant to take us with a baby. This worried us because we had to get somewhere to live, and we pleaded our case. However, she said she would take a chance. It was difficult to get rooms when there was a young baby involved and we had a number of refusals before we tried this house. We had scoured all the 'Room to rent' addresses advertised in various shops before we got lucky here. We stayed there until we bought a house in Arlington Road, West Ealing, W13 after about 6 months. While we were in Hanger Lane, the landlady used to watch for any pram marks on the walls, and floor, and kept a tight rein on the lodgers. No noise was permitted. The room was about ten feet by twelve but the paintwork was not great. The ceiling was papered, and two sections were starting to peel off. A gas fire, and small gas cooking hob were our means of heating, and cooking. We had to feed the gas meter on a regular basis.

Next to our room there was another room, occupied by a dodgy character, who was big, fat, and slovenly looking. When

I went off to work in the morning Molly kept the bedroom door locked, and usually waited until this guy left the house, before she went for a walk with Ann. One evening when I came home Molly was in a dreadful state, frightened, and tearful. She told me that when I left for work in the morning, she had Ann in the bed with her, and of course Ann was jumping around, causing the headboard to bang against the wall of the adjoining room. He took this as a signal from poor Molly that 'the coast is clear, I'm waiting.' He immediately tried to get into our room, and Molly told him to go away. He persisted, and eventually the landlady heard her cries for help. She immediately gave him notice to get out of the house. After that we became great friends with the landlady.

One Saturday morning we were in bed, and as usual Ann was in the cot next to us. I don't know what woke us, but Ann had been jumping around in the cot, and suddenly she went quiet. I looked at the cot, and there was no sign of Ann, accept a pair of tiny legs wriggling, and sticking up between the cot, and our bed. She had been trying to get into our bed, and fell between the bed, and cot, and her face was virtually stuck into our mattress. All was well, and she was none the worse for her escapade.

We had been actively looking for a house to buy, and eventually we found our first home at 24, Arlington Road, in West Ealing, W13 and we paid around £4,000 for it. I had a fair deposit accumulated from my time at sea so we did not have too large a mortgage to repay. However my job as Electronics inspector at E.M.I. Hayes in Middlesex was not great so we had to supplement our income some way. My pay was £11.00 per week. In fact I was paid by the hour and I remember I was given the usual one penny an hour rise.

The layout of the house comprised, four bedrooms upstairs, and downstairs we had a sitting room, which we converted to our double bedroom, and where later Sean, and Jane were born. We had another room, plus a dining room, and finally a kitchen. There was also an outside toilet. We took in lodgers, supplied them with bed, linen, and bathing facilities.

We also washed the linen. We had to get metered heating, and cooking facilities, installed in each room we let out. Sometimes we could have up to four people lodging with us, including married couples, or two girls to a room. It was a hair-raising experience to encounter the carryings on with some of these people. These are a few examples.

One old lady, who had been put out of another house because of her age, was delighted to come, and stay with us. However, we had a rule that the downstairs was ours, and no lodger was to encroach there. Over a period, when the old lady had been with us for a while, she felt like one of the family, and took a shine to Ann. This led her into the kitchen, and out into the back garden. Eventually she wore us down, and we got quite friendly towards her. Later, when we had to sell the house, it was heart-rending for Molly to let her go. The old dear cried, sobbed, and wondered where she was going to stay. Since she had plenty of notice, she did eventually find a place.

We had a fellow who was a runner for the bank. He was spick, and span. Well groomed is an expression that falls well on him. He would have been in his mid thirties, and never seemed to have friends, men or women, and he said that he was unmarried. All in all he was one of the better lodgers.

We had a big fat man who was an engineer. Again he paid his rent, and did not cause us any grief, but we were suspicious of him. He did not appear to have any male or female friends and he never mentioned any, but Molly would not let Ann, Sean or Jane near him. Moll's sister, Anne, was in the house one day. Anne said to Molly, "Don't worry about him, he's neuter gender."

There was another lodger, a woman whom Molly christened 'The Cougher.' The woman was in her forties, and she was a heavy smoker. This was the cause of the constant coughing. She also had another annoying habit of coming out of the bathroom with powder all over her feet, and walking on the stair's carpet to her bedroom. She never wore slippers or cleaned up her feet marks. Eventually, we had to give her notice to quit, because of this, and the coughing at night.

In the front double bedroom we had a wonderful young couple that stayed for about a year. They were class, and absolutely no trouble. This was in complete contrast to the two young Irish girls who took the room after them. We had nothing but trouble from these. The first sign was the loud music, which we had to curb on a number of occasions. When threatened with notice they of course knew all the ropes, and it was easy to figure out we would be tested to the limit. I started to make notes of times, dates, and warnings, etc. to get rid of these girls. On one occasion I found that they had removed the landing bulb, and inserted the plug of their iron into the socket. They hoped to save paying for their electricity by doing this, as the lights on the landings were our responsibility. The stress on the iron pulled the wires from the ceiling rose, and I had to repair the damage.

We had a rule that no boyfriends were allowed. One night at about ten-thirty p.m I was passing by the bathroom, and heard grunting, and whispering. I immediately became suspicious, called Molly, and we waited out of sight at the bottom of the stairs, and near the light switch. Eventually, the bathroom door opened, and one of the girls appeared, and looked around. She then signalled, and a young buck came out, and started to come down the stairs, careful not to make any noise. When he was halfway down, I switched on the light, and accosted him. The expression of disbelief, and fear, in his eyes was something to see. He stood transfixed. I threatened to have him arrested, for breaking and entering. He spluttered that the girl invited him. Great, now I had the evidence I wanted, and told him to get to hell out of the house. I never saw anybody move so fast. The girl was watching pale faced, at what was happening, and ran upstairs when she saw me coming up. I knocked on the door, of the bedroom and told the girls they had to be out by nine a.m. the following morning. They threatened to take me to the rent tribunal, as they were entitled to a week's notice. I said no way were they going to stay here for a week, and that if they refused I would get the police, and tell them what had happened. They were gone by eight-thirty a.m. next morning.

There was the case of the 'Fly by night couple.' We rented a room to a couple who were in their thirties. After a few months with us, Molly and I were browsing around one Saturday morning, and became suspicious, because unusually for this couple there was no sound from their room. I went up and knocked on the door to check if all was well. There was no answer. I tried the door, and it opened. Surprise, surprise, and swears. The couple had flown the coup. They had sneaked out while we were watching TV the night before. Everything of theirs was gone, and so were the keys of the bedroom, and more importantly the front door keys were missing. On the following Monday, Molly went to shop just down the road, and lo, and behold, who did she see inside the shop but the fellow who had left with our keys, and rent. She went up to him, and he got the shock of his life as Molly demanded the keys, and rent. He brushed past her, but she saw him get into a van with a well-known brand name on it. When I got home I checked the phone number, and rang the firm, and asked to speak to this man. He was out, so I left my name, and asked them to get him to return the keys, and rent. We got both.

In the single bedroom we had a young girl who nearly started a fire. I smelled smoke one night, and traced it to this room. I banged on the door, and a sleepy eyed girl opened the door. She had been asleep, and left a chair too close to the electric fire. The chair was smouldering away in the room, and she slept through it.

We rented a second single bedroom to a man who worked nights. He usually left early in the evening, and came home before we got up. It took us a long time to find out that this lodger was not only working nights, but that his girl friend used the room during the night. Before going to work he would smuggle her into the room, and she would leave as he arrived back from work. He expected two in residence for the rent of one person. How long, or how this got by us I'll never know. We learned a lot in this house.

One morning Molly and I were going for a stroll, and we had Ann in the pram. As I opened our front door, three firemen burst through,

"Where's the fire?" they shouted, and they nearly knocked me over.

Outside, on the road was a fire tender, with more firemen and all were looking into the house.

I thought, "Am I missing something here?"

Ann started to cry, and Molly was asking what fire they were looking for.

One fireman said, "Is this 23?"

I said "No, that's next door."

They apologised as they left, to put out a chimney fire at No 23.

Our neighbours were a mixture of nice, and some nosey people. One woman who lived about six doors away saw a new Hoover twin tub washing machine being delivered to us, and next day was down with a pram full of dirty washing, to know if we would let her use the machine. I politely said she could not use the machine, and this upset her. Big deal! There was a launderette down the road, and I told her to go there.

We had very good friends like the Irish couple Meta, and Sean Purcell, who lived on another adjacent street. There were also our next-door neighbours, Miss Godfrey, and her sister, a lovely genteel English family of two elderly unmarried sisters. They were Protestants, but despite this the Catholic Archbishop of Westminster came to visit one of the two sisters, and the press was there. I know she had been very active in some charity work, and had some award presented to her. One sister used to invite Molly in for a Sherry a couple of times a week.

Behind us, in Waldeck road, we had great friends in Pat, and John Greene. Molly and Pat used to jump over the wall at the end of the garden, and visit each other for a chat. One night, Molly returned over the wall from Pat's house. They had been sitting in front of at big warm fire, having a tot of whatever, and Molly began to feel queasy, so she decided to come home. Molly just made it to the outside toilet, where she got sick into the bowl. She flushed the toilet, and got up feeling great until she left a yelp, and exclaimed, "Jesus, my teeth" She had lost them when getting sick, and she flushed them down the toilet. She never lived that down. She used to keep her hand to her mouth while talking, until she got a new set of dentures. Pat Greene went into hysterics, laughing.

During the years in Arlington Road we were very happy, and soon Sean Kevin was born. His birth was difficult for Molly, and Sean. When delivery time came a Greek midwife attended, and would not call the doctor insisting she could manage. A delivery by her, without a doctor in attendance, meant more money for her. Molly was struggling badly trying to deliver Sean, and I was with her holding her hand, and giving her the mask. Eventually, the nurse got really worried when she discovered the chord was around Sean's neck strangling him. She called for the doctor, but before he could come, Sean was delivered. He frightened me when I saw him. He was like a skinned rabbit. The poor child was blue and seemed to take ages to start reviving. The chord was wrapped around his neck stifling off the oxygen. The bitch of a nurse nearly cost us the life of our only son. Molly and I have never forgotten this woman. Thank God, there was no lasting damage done mentally or physically to Sean. I picked him up kissed him, and welcomed him into the home, having made eye to eye contact. Sean grew up there and went to Drayton Green School, until we came back to Ireland. However, there were to be a number of worrying times with Sean.

The first was when he had high temperatures, and used to go into convulsions because of this. One day, I had him in my arms, out in the garden, on a summer's day, and a plane flew overhead. I looked up, and as I did I felt Sean go rigid in my arms. Fortunately his Aunt, Clodagh, who was Paddy's wife, and who was a nurse, was with us. She made me take him indoors, as he was having a convulsion. We laid him on the bed, and stripped the clothes off, as his temperature was way up. Clodagh saw that Sean was going blue in the face, and she put his head to one side, as he was choking on his tongue. She put her fingers into his mouth, and cleared his tongue. Without her would we have coped? I don't know if we would have been alert enough, due to worry, but I'd like to think that calm would have prevailed. Thanks Clo!

The next frightening episode with Sean was again high temperature, and the doctor could not figure what was wrong, so he asked me to take Sean to King Edward Hospital. Since I did not have a car and I did not have a phone to call a taxi, I took Sean in my arms on the bus. I felt this was quicker than looking on the main street for a taxi, and then going back home for Sean. Sean was in a semi trance at the time. Eventually, a doctor came out to the waiting room in the hospital, and asked various questions about symptoms.

"Did his bowls move," the doctor asked.

I said "No" as Sean had started crying due to all the prodding.

The doctor then left and I was alone holding Sean.

"Stop that baby crying." One nurse blurted out.

I lost my temper, and told her to learn tolerance, and understanding. Sean was taken for X-rays, and I don't know what else they did, but they brought him back to me, and asked me to wait while they checked the test results. Without warning Sean let his bowls open all over me. It came through his nappy, and was down my front, and my sports jacket was covered in a greenish/yellow watery mess. The smell was unbelievable. I gagged. They took Sean immediately, and said they were transferring him to the Middlesex Fever hospital. He was apparently suffering from severe gastroenteritis, and would be in hospital for quite a while. Nobody gave me even a hint that I might become infected.

I had now got to get back home, smelling like a sewer, and did not relish getting on the bus, but with Sean on my mind, the smell became academic. Everybody I passed gave me a wide berth, and terrible looks. When I got to the house Molly was waiting for news. She had to stay behind with Ann. She held her nose, while I recited emotionally what had happened in the hospital. We both cried for quite a while. My clothes went into the bin.

We went and visited Sean in the fever hospital every day. It was great to see him improving. One thing that always stands out is the greeting we got from him when he saw us. He was strapped in his cot, but had plenty of freedom to move, and jump around. At first he used to cry a lot when we left, but we soon found he had other interests which occupied him and made him forget us, or that we were even there. When he heard the food trolley coming he would go into raptures, and had eyes only for the door, awaiting the food. We were non-existent at this point in time, and it was a great and easy way to say our goodbyes without upsetting him. When the food came in, he grabbed the plate and went to one corner of the cot, sat down, and made short work of any dish put in front of him. Eventually, all the tests came back clear, and he was allowed to come home.

Sean got over this setback but on another occasion, when I was returning from work one evening, I was motor biking up Avenue Road, in West Ealing when I saw Molly with Ann, Sean, and Jane. Sean's face and forehead were a mess. He was black, and blue, but in good spirits as Molly had just bought him an ice cream in the nearby shop. Molly explained that she had been window-shopping at a large store, and Jane was in the pram, while Sean was seated on a seat fitted to the pram. Ann was holding Molly's hand. For a little while Molly took her hand off the pram, and it toppled forward. Sean hit the ground, face first, hence the damage to his face. A woman nearby helped, to get Sean to a doctor who gave Sean the all clear. Poor Molly was very upset and scared that Sean might have been seriously injured.

After Sean, Jane Caroline arrived. As Molly's time drew close we were in the bedroom, and Molly began to get the pains very quickly. I knew the time was near, and that it was time to get the doctor. As we had no phone in the house, and the nearest public telephone was about a quarter of a mile away, I was in a quandary what to do. I did not want to leave Molly alone, and I had to consider Ann and Sean. It was two a.m. with nobody to help. Neither Molly, nor I thought of asking Pat Greene to pop over while I went to the phone. I could not believe how badly

I had prepared for any of the births. My conscience began to explode with guilt.

There was no option but to go to the phone, so I went outside the house, looked up and down the street to see if there might some help around. My prayers were heard, and I saw a bicycle light approaching along the road. I shouted for help, and this young girl got off her bike. When I asked her if she would go and phone the doctor for me she asked what the problem was. When I told her the reason, she smiled, and said "I'm the local district nurse, lead the way." I was never so relieved. She met Molly, and they got on like a house on fire. The nurse was from Mallow, Co. Cork, so they had a lot to chat about. I got the doctor, and the nurse kept things ticking over saying Molly was not in immediate danger of delivering. I remember poor Molly haemorrhaged badly, despite the best efforts of Dr. Wrangham. After delivery she lay there on top of the bed, shivering. The floor was covered with newspapers, stained with blood. Jane was washed, and put into warm blankets in a drawer on the floor. When I went over to her she had skin like peaches, and her two hands were in front of her face, and I swear she was actually looking at them, and flexing the fingers. I witnessed this delivery as well as I had Sean's.

Like Ann and Sean, I kissed her and welcomed her into the home. One thing I always think about was, as Jane grew up, but while still in nappies, and not yet able to walk, she took off across the road on her bum, pulling herself with her right hand, while she pushed with her left hand, which was between her legs. Boy, could she move. She was across the road, before I could get her up in my arms. Later, she liked getting a spin on my, 250cc BSA motorbike. I took her around the block, and she loved it. Later, in life she used to ride motorbikes. It was an absolute joy, and difficult to express how lucky we felt having three beautiful and healthy children. Everything was in order, and in the correct places, in each of the children. Thank God we were so lucky.

While in Arlington Road, my first job was at E.M.I (Electrical and Music Industries) in Hayes, Middlesex. I worked in the commercial computer manufacturing side, before being transferred to the Military areas, which were top secret. I worked on the side ways looking radar for the TSR2, Swing wing Bomber, and on the 'Blue Streak,' missile. It was here where I got my first sight and knowledge of semi-conductors, transistors, and diodes, plus various new technical advances in electronics. I really enjoyed the work here, but the pay was not great, so I decided to change jobs.

During my time at E.M.I I bought a motor cycle of early vintage, and had been converted from a solid front fork, to a telescopic type fork. It was a 250cc BSA Enfield. My first attempt to sit on it, not to mind ride it, nearly ended in damage to the seller's front gate. The bike was parked on a sloping drive and I got on it. The stand lifted and the bike headed for the gate with me sitting on it. The owner shouted, "Brake! Put on the brakes!" I stopped the bike about two feet from the gate, and I was shaking like a leaf. I got off the bike, and decided to buy it even though I had never ridden a bike before. The next day, my neighbour and friend, John, who owned a motorbike with side car, came with me and we collected the bike which he drove, with me as a very nervous passenger. I practised riding the bike around the streets of Ealing and gradually got to get the feel for it before I attempted to go to work on it.

One morning I was on my way to work, riding my motorcycle, on a frosty ice covered road, when I came off the bike with a thump. I was not too happy but fortunately I was wearing my helmet and the proper bike gear. It happened as I was going through Southall in Middlesex and I heard a loud noise overhead. I glanced up to see a plane flying very low, coming in the land at Heathrow. At the same time I saw the red brake lights of the van in front of me. I hit the brakes and the bike went under the van as I hit the road and came with a bump against the kerb. Fortunately all I suffered was hurt pride and duly embarrassed I immediately got up, mounted the bike and went to work. To avoid a ribbing I didn't tell any of the guys at work.

On another occasion I was about four miles from home and got a puncture. For about one mile I pushed the bike and was feeling knackered. A garage loomed in front of me and I bought a tin of puncture repair fluid. The instructions were to pour some into the tube and it would seal the puncture. I tried. I pumped the wheel to no avail. Eventually I poured all the liquid into the tube and no good. There was nothing for it but to push the bike home. After about another mile I gave up and started the engine, mounted the bike and rode home, swearing and feeling browned off. It took me a week travelling by train before the bike was on the road again with a new tyre and tube.

One day when I was at work I got a telegram saying that mammy was not well and to come home quickly. One of my friends arranged for a ticket on Aer Lingus for me and offered to drive me to the airport. I met him at my home where I packed and was not feeling too good as I hoped that mammy was not too far gone, and would be there for me to say my good bye. On the plane I was agitated and unsettled hoping the plane would move faster. We hit turbulence and were shaken up a bit about ten minutes out of Dublin. Eventually, I got home and Sheila, my sister, met me at the front door and told me that mammy had died a few hours earlier. It was at the time we hit the turbulence on the flight over. I went immediately where mammy was laid out in the bed. Daddy had told everybody to give me space and to let me be with alone with mammy I was thankful for that as I cried my eyes out looking at my dear mother's lovely face. One eye was slightly open and I tried to gently close it. I kissed her cold cheek and wept some more. Later, I carried her coffin and was glad I had at least got this to soothe my loss. She had died four days after her 61st birthday. She was so young. I still miss her and remember all the laughter and good times we had together

For some time I had been looking for better opportunities to improve my career and to get more experience so when I saw an advertisement for a job that appealed to me I immediately applied. From EMI I went to work as a Quality Engineer, for Standard Telephone and Cable Company Ltd. in New Southgate, North Circular Road in London. It was a nice motorbike ride along the N.C. road every day, and was much easier than public transport. Here I was employed as Quality Engineer, for more Military equipment. I worked on Walkie Talkies, for the German Army, under very tight quality criteria. Due to a large number of rejections by the Germans, I was temporarily transferred to S.T.C in Newport in Wales, to sort out the problems on the factory floor. I was successful in this and got the acceptance rates up to the required level. It also meant I was away from Molly, and the children from Monday, to Friday, for about a month. Whilst I was getting extra money in the form of allowances, I was glad to get back.

Outside of work we had other obligations and amusements that we looked forward to on a weekly basis or more often if the occasion arose. We attended Ealing Abbey for mass. This beautiful Abbey had been badly damaged during the war, and part of it was screened off from view. I remember the priest was looking for donations to rebuild and reconstruct the damaged sections and I like others volunteered to contribute, by making a covenant for life, or otherwise revoked by the church. The priest stated that anyone paying taxes could avail of this option, and that it would not cost anything, except to the tax office, that would refund the tax collected to the church, as a charitable donation. This went fine until the end of the tax year, and I got a bill from the tax office to refund this money to them, as I had not earned enough to pay tax. I went to the priest, and showed him the letter and he immediately revoked my covenant, and said he would repay the money I had paid. This money he said was to, 'come from, the following church collection plate.' He was true his word.

It was here in Ealing that I bought two lovely antique clocks for a knock down price. An old man had collected lots of antiques, paintings, statues, and clocks during the war, and had them in his house in West Ealing. I went there one day, and was flabbergasted with the amount, and variety of clocks, from mantle clocks to chiming grandfather, and grandmother clocks from various countries. He told me I could have any of them since I seemed so interested, and he was too old to take care of them. He wanted them to have a good home. I choose two.

One was an English kitchen wall call clock, circa 1803 made by "Gregory, Basingstoke,' and the other was a German chiming clock.

    Working in England was bearable but both Molly and I wanted to return home to Ireland so I could not believe it when I found out that The General Electric Company of America was opening a factory in Ireland and that they were hiring personnel. Molly and I decided I should apply. Duly elated, I wrote and enclosed my C.V and soon after was called for interview in Dundalk, Co. Louth. When the date for me to go arrived I hugged Molly and the children and with my suitcase packed I headed off via Aer Lingus for the most important interview I'd ever had, which would hopefully be my passport back home.

Chapter Thirty-one

On arrival in Dundalk, I was interviewed by the Director of Quality, who was American. With my background in electronics and particularly in semi conductors this man felt that this is where I would be best suited and he wanted people such as me to help in the establishment of a semi conductor factory. When he asked me if I would like to start soon I told him I had to give in notice to STC and would love to join the 'ECCO,' Company, which was a subsidiary of General Electric of America. This company was bringing in new technologies, to this country. They manufactured Diodes, and Transistors. The processes included all states of assembly, inspection and tests. I was employed as Quality Specialist, and had numerous Inspectors, and Process checkers, reporting to me. I grew to love the job, but being in Dundalk in 1969 had its own problems.

When my notice with STC expired I came in advance to Dundalk in order to see how things were, and what the housing situation was like. I took up lodgings in a house, opposite the Garda station, and in fact a Sergeant and a Garda lodged in the same digs. There were a number of Mormons also lodging there but at no time did they try to discuss religion. The house was subdivided with rooms partitioned, so that the owners could let out as much space as possible. I can't say I liked the place but lodgings were hard to find at the time. During my time there I developed severe bi-lateral septic throat, and had to stay in bed for a week. My bed was soaking in perspiration, and the landlady did not call in once to find out my state of health, or to change my bedclothes. All she wanted was my rent. The doctor gave me a double shot of penicillin so that I could go over to Molly and the children for Christmas. Molly was shocked when she saw me. I had lost a lot of weight, and looked gaunt. When I returned to Dundalk I was much stronger, and waited patiently for Molly and the children to join me.

Despite my joy at the idea of coming back to Ireland to work, it was not easy on Molly, or the children, who stayed in Ealing, until we decided it was safe to sell, and for all to come over. Molly was having her hands full doing everything over in Ealing, and with no help from me. What a girl I married. When the chips were down she coped with all that could be thrown up. Eventually, we decided to sell, and fair dues to her Molly got on with the job, and sold the house. She took care of all the legal aspects and eventually, got tickets to come over to Ireland. My motorbike was left in the front garden and the paperwork for it as left in the house. I wonder what happened to it. It brought back memories of my brother Anthony when he had the accident coming to my wedding and he too left his motorbike in somebody's front garden.

I was over the moon when my family arrived at Dublin Airport and we hugged, laughed and generally went wild. It was great. However, our joy was put on hold for awhile. Since we did not have anywhere to live together, Molly moved in with my sister Sheila, and her family in St. Martin's Park, in Rathmines, Dublin. Bennie, and Sheila were very good, and I shall be eternally grateful to them for the help and comfort they showed to my family. Molly was there for months. I used to go down at weekends, until we located a house in Dundalk. Molly liked the house, but it would take some time before we could move in. We therefore decided that Molly, and the children should move to Cobh, where Mammy and Daddy welcomed them with open arms.

Eventually, we settled at 'St. Monica's' in Long Avenue, Dundalk, also known as Avenue Road. This was a four bedroom detached house, completely walled in and bounded on one side by the ramparts, which is a stream The other walls were eight feet high and we were very secluded. There was grass all round the house, and this kept me busy with the lawnmower. We soon settled in, but found that we needed to replace all the rotting sash windows, and also found woodworm in the attic, that had to be treated. There was an Aga cooker in the kitchen, and this kept the place cosy and warm. The next thing was to get a

TV, so we had an aerial erected, and bought a twenty-one inch Ferguson. I bought a white, second hand 'Fiat eleven hundred,' from a Dublin Garage. It was a bad buy. It had poor headlights and the handbrake used to freeze up in winter and could not be released for a long time. Later, when I was driving to Cobh, with the whole family, because Daddy was dying, the car broke down at the Coachman's Inn, just outside Dublin Airport, where I had to get assistance from some of the customers to push start the car. Later, I got a speeding ticket going through Abbeyleix where I was allegedly doing thirty two mph in a thirty mile mph zone. The exhaust pipe burned out, and one of the lads at, ECCO later replaced it with a copper pipe. The aluminium engine was a washout, and I'm sure the clock had been turned back. I suppose going from Dundalk to Cobh did not help particularly in those days of bad roads.

In Cobh, I went in to see daddy but to my dismay and sorrow he had died. Apparently he had a massive heart attack and went quickly. My sister Nora was with him and arranged to have him laid out in the back room behind the shop area. I was very dismayed and upset that I had been late again but once more I had the comfort of knowing that I carried his coffin to lay him to rest next to mammy. When I had looked at his corpse his face looked as if he had suffered quite a bit during his last few hours. The sciatica which he had complained about in his left arm was in fact the classic sign of heart problems but he was unaware of this. His last moments were traumatic for Nora as he struggled to get his breath during a massive heart attack. I wondered what now lay ahead as both my parents were dead, but I had my own family to look after and intended to do that to the best of my ability.

When we arrived back in Dundalk I knew I had to get rid of the Fiat which was clapped out. Each day when we went out I kept an eye out for a good buy and one day while out with Molly and the children, we passed a garage and there was a grey Cortina for sale. I free-wheeled the Fiat in, and ignoring the exhaust noise, the car otherwise looked spotless and in great shape. I asked the garage owner how much he would give in a trade in. I had looked over the Cortina, and she looked fine, with not too many miles on it. We came to a deal, which I felt I had got the better of, and I came back with a cheque, took over the Cortina, and drove home, delighted with myself. On three occasions I had calls from the garage owner who had sold my Fiat on, and had it returned to him as 'poor value and not roadworthy.' He wanted me to compensate him' but I told him he had made a deal' and took the risk as I had. I heard no more' and then one day I saw the Fiat trundling up the road, and an old man quite happily driving it, at about twenty-five m.p.h. He waved as he went by.

It was here in Dundalk that Molly got the shock of her life, when she found she was pregnant again. She had to have it confirmed and went to her doctor named, a man named O'Reilly who hailed from Cork and who practised in Dundalk, but was only licensed to do deliveries in Newry, Co. Down. One evening he knocked on the door after the tests had been done. I opened the door. Molly stood alongside me, worrying, and before anybody could speak, Ann piped up,

"Is mammy going to have a new baby?"

We stood still as Dr. O'Reilly smiled, looked at Ann. and said,

"Yes she is. Daddy goofed again."

After a while Molly got used to the idea, and when her time came we went to John of God Hospital in Newry, Co. Down and Susan was born there.

She had dark hair, and kept sticking her tongue out. Molly said to me,

"Is she alright; is there something wrong when she keeps sticking her tongue out?" I picked Susan up, kissed her, and again welcomed her into our ever increasing family. I eventually set Molly's mind at rest, when I asked the Doctor to explain that there was nothing abnormal with Susan. Molly used to call her "My dark haired Susan." We returned with a family of four children; one Corkonian, two Londoners, and one born in the North of Ireland. It sounds like we're a bunch of itinerants! Susan was a pert little baby, and her hair grew blonder with time. It was lovely to see all the children playing around in the large garden, and they had many friends.

Around this time the troubles started up in the North of Ireland. Molly was getting quite concerned, because Dundalk was a known area for IRA activity, and she feared that the troubles would spread down south. Since we were adjacent to the main Dublin road, her fears gradually grew.

"What will we do if 'THEY come down here?" She used to say half jokingly.

To make matters worse, our house was next-door to 'Failte House,' a guesthouse, which was to inadvertently cause Molly more worry. Right smack in the middle of the troubles the owners, our friends, used to take precautions to lock up, and make sure that no subversives from any side could get into the house without their knowledge. The owner's wife told Molly that one night a man, and woman came, and booked in, and the owner locked the doors as he usually did. They then retired, and the next morning the guests checked out. Later, on the TV news, a picture of a well-known IRA man on the run appeared on the screen. They looked dumbfounded at the TV, as it dawned on them that their overnight lodger was none other than this 'Fugitive'. They realized then, that rather than locking out undesirables they had inadvertently locked one in. As I said, Molly was not too impressed.

The time came for another move, and this time it was going to be Dublin, where we felt that the children would have better educational facilities, and a better way of life. We too, would benefit from a move. While in Dundalk, despite having very good friends, and workmates, and liking the town I was now around forty, and felt I had to make the move now. Molly and I decided after some searching, and visits around estate agents, that Malahide would be a perfect place to buy a house, and plant our feet. We looked and fell in love with plans for a small estate at Gaybrook Lawns, and took up an option to buy before the house was built. Again, it was Molly who found the location, while she was with another friend from Dundalk, who was also on the lookout for a house.

We had to break the news to the children that we would be moving and leaving Dundalk. After we told them, Ann and Sean sulked and went out in the garden where they had the following Pow-Wow;

Ann: "They don't care about us, they only think of themselves."

Sean: "Yeah and what about all our friends, we won't see them again."

Ann: "They're selfish, and I want to stay here."

Sean: "I do too, and I don't want to go to Dublin."

There was more muttering, but Molly and I could not quite make it out, as they moved away from the window where we were listening and trying to control our laughing. Eventually, they got the message that we were going, after we told them they could stay, and we would go alone.

At work I went into my boss's office to hand in my notice. Tom Cullen and I got on very well from day one, and I was sorry to be saying goodbye to him. The conversation went like this,

"Tom, I want to hand in my notice, and will be leaving in a month."I said.

"What brought this on? Are you not happy here?" asked Tom.

"I am Tom, but for future prospects, and for my family, I feel its right to go now."

"Is there nothing I can say to get you to change your mind, and stay?" he asked.

I answered, "No, Tom my mind is made up, and I have already accepted a job elsewhere."

Tom got up from his desk, and held out his hand,

"I'm sorry to see you leave Jack, but before you go is there anything you would like to say about me, and my style of management?"He asked me.

"Tom I was very happy to work with you, you have been very fair, but I have one criticism and that is whenever you gave out a task. You did not follow up to find out if it had been completed." I replied.

Tom shook my hand he said, "Thanks Jack for being so forthright, I'll remember that in future, and on behalf of ECCO I wish you all the best in your new job." He then sat down.

Within a few seconds he was on his feet again, put out his hand, shook mine again and said,

"Welcome to Core Memories I'm your new boss there."

He too, was on his way to my new Company, and was in fact the one who vetted, and selected me, for my job. I nearly fell over, and he laughed his head off. For nine months, Tom and I drove from Dundalk, to Dublin every working day, before we eventually moved into our new houses. It was tough going. Tom went to live in Sutton. Incidentally the lady who was with Molly searching for a new house was Tom's wife. More friends kept arriving from ECCO, and it was becoming 'home from home.'

During my daily trips from ECCO to Dublin, I used to call into our estate, to monitor progress on the house, and to push for completion. The foreman was sick and tired of me, but I didn't care, and in March, 1970, Molly and I took possession of the house. Molly was thrilled, and started organising the furnishings, and decorations. We spent lots of time selecting carpets, curtains etc. However, we were driven up the wall with the construction going on, and since we were the second family to move in there was a lot of work to be carried out by the builder. There was dust, and muck everywhere, and the children had a whale of a time.

Gradually, families began to move in, and plenty of young children started to arrive. We had some great neighbours, and many houses changed hands over the years we still kept some contact with many of them.

I continued to work for Core Memories, which was a labour intensive operation. It produced cores for memory systems, and also produced core memory stacks and systems for computers. These were the state of the art until micro chips entered production. The company was later taken over by Data Products, which saw out the end of core production, and produced Line Printers for the Banks and large companies, like Prime, Honeywell, Siemens, Olivetti and others. I gradually got promotion, and became Quality Control Manager and later Production Manager of some products. I was also

Project Manager for the final successful completion of the semi automated production line for printers, paint shop and associated parts throughout the plant. The system was installed by a Japanese company. Later, I became unit Manager of one of the two manufacturing facilities. In this capacity I had a lot of responsibility, and took it upon myself to do a lot of weekend work. Looking back, I was a fool that I did not delegate more work. I was elected as the shop floor representative on the pension's board, and appointed Secretary of this board.

The children in the meantime had started school in Malahide. Ann went to the primary, 'Red School' at Yellow Walls road. She then went on to secondary at Scoil Iosa. Sean attended St. Sylvester's primary school, and then to secondary at Christian Bros. at Colaiste Mhuire, Swords, Co. Dublin. Like Ann, Jane and Susan also attended both these primary, and secondary schools.

Molly always made sure that the children were clean, and neat, going to school and she took more interest in their friends than I did, due to the long hours I put in at work. She was the one who got the school results, and complaints, if any, and I never got to hear about them, until the children were grown up, and could laugh and say, "Dad, you did not know half of what went on." I now wonder if they will wonder what went on in my life when they read this! God bless them all there was nothing malicious in what they did, and I can honestly say I'm proud each one of them, and what they went on to achieve. Many people complemented Molly and me, saying what a wonderful family we have. Unfortunately, Molly is gone, having done all the hard work. Well done Kids! Another thing, could I have given out to the kids having looked back at my own youth and life? The proof of the pudding is in the eating; each of them is married, and owns their own home. Their homes are a credit to them and we have 6 wonderful grandchildren….

Other parts of our lives complimented the rearing of our children and Molly and I took all opportunities to do what we liked to do. One Thursday in Malahide, Molly and I decided to go to Denis Drum's auction, and as I had to park the car

Molly went on in front of me. When I got to the auction room the auction was in full swing, and I could not see Molly due to the packed hall. I waited for our 'piece' to come up, and we had decided to bid up to a certain amount, and then stop. All was going well as I bid, only to find I was reaching the critical bidding stage, which we had agreed on. Our top price was reached, and I decided that since Molly wanted the piece I would continue bidding to try and get it. I succeeded after going a few quid over the price we had agreed, and feeling very cocky, I met Molly after the auction. When I told her about my achievement she was not at happy, as she had been the only other person bidding, and we could have had the item for a lot less. Later we laughed.

Molly, and I used to take the family to Cork regularly when they were young, and it used to amuse me watching Molly getting excited, as we got close to Glanmire - on the outskirts of Cork City - out came the comb and handkerchief to clean the children. This was a ritual. No way were her children going to look out of place and Molly was so proud of them.

We had many holidays in England, California, Ibiza, Portugal, and in Italy. Besides our honeymoon in Germany, our first holiday was in Majorca. When we landed there the first thing Molly did was to look around her, and exclaimed, "What am I doing here? I should be at home with my children." Molly always thought of others before herself. Anyway, she got to like the place, and we had some great times there. We also went to the Algarve in Portugal. She was not too happy there, because there was a lot of building going on, and it was quite dusty and noisy.

Once, Molly and I set out for a holiday in Cork, and Kerry, with Paddy and his wife Clo. We were halfway to Cork when Molly asked me if I had put her stuff in the boot. I told her I assumed she had done so as she still had it when I put my suitcase in the car. Panic set in.

"You're joking, you did pack it for me, didn't you," she exclaimed.

"No, I didn't," I said, as we pulled over on the side of the road, where Clo and Paddy also pulled up in their car. They were wondering what was going on, when they saw Molly frantically opening the boot. The colour drained from her cheeks. She again searched the boot of the car and of course there was nothing to comfort her.

In a bewildered voice, she exclaimed "I have nothing to wear and no change of any clothes." Gradually, we all started to laugh, and poor Molly had to watch as we were doubled over. She took it fine, and borrowed some clothes from her sister Eileen, in Mallow, and bought some new underwear. She got on fine, and we had a lovely holiday in Creman, and around the ring of Kerry. In the midst of all this happiness we were in for a rude and painful experience which was to shatter our long term dreams....

Chapter Thirty-two

Over the years Molly had her traumas with deaths in her family. Her mother Anne died on January 31st, 1984, aged eighty four. Molly was at her bedside when she died and it naturally upset Molls for quite a while. Next to go was her twin brother Paddy. He died of cancer, on February 21st 1990, aged fifty eight and Molly was extremely upset when Paddy died. I suppose it was the twin relationship, and Molly kept playing tapes of Spanish South American music and tangos that Paddy had given her. One particular tape that I remember was the South American Group called Los Fabulosos Paraguayos. She played their music day after day in the kitchen, and I tried to console her but she sobbed, and it took a long time for her to get over this death. Doreen the eldest, died sometime around 1993, and cancer was again the killer. However, Doreen had been a heavy smoker and after the death of her husband Leo, she became very lonesome and kept to herself. Molly went to England for both funerals of Paddy and Doreen.

Poor Molly then had to contend with my illness when I developed a heart problem and had to have Bi-pass heart surgery in 1989, Molly seemed to be stressed. I don't know if this was the trigger for her developing cancer in her oesophagus over the years that followed. I know the main cause was cigarette smoking. It took a very long time to confirm the diagnosis. Molly began suffering and had difficulty swallowing food. She had to attend a consultant who performed a throat stretching operation under anaesthetic. I drove her regularly for this to be done as a day patient, and waited until she was wheeled out. She told me it was a terrible ordeal, with this metal tube inserted into her throat while she was still semi conscious. How I felt for my poor darling, but she got some respite for a short period, before it all had to be done again. One horrible day she was told that a biopsy had shown a malignant symptom, and she would need drastic surgery to remove the oesophagus, and this meant that they had to rejoin her stomach, directly with her throat.

This was done by opening the neck, and stomach, and stitching together what was left of her throat minus, the oesophagus. She was then stapled at the wounds. My poor darling suffered a lot. It was felt that the cancer had been contained, and had not spread. How wrong can you be? After various X-rays, and further tests, good news continued to boost Molly, even though she was not progressing as well as she should have. We had been to Italy for Sean's wedding, and later on we went over again for another visit while she was ill with cancer. She did not enjoy this visit as she was in pain, and I was too dumb to take full heed of the seriousness of her condition.

One day a family friend, Molls brother-in-law, were at our house enjoying a holiday. He was a doctor and did not like what was going on, and asked Molly to come to Cork where he would have her checked out. She was booked into hospital there at Shanakiel hospital, and the first X-rays indicated widespread secondary cancer. There was no hope. Neither Radio Therapy nor Chemotherapy was recommended. What a devastating blow to all of our family. Grief was flooding through me as I brought Molly home. All the fight was gone, and she was tired out. All we could do was wait, and make Molly as comfortable as possible.

After a few months Molly had to stay in bed and Nurses from St. Francis Hospice paid daily visits. They eventually had to give morphine. These nurses are a credit to their profession, and never failed to administer love, and help to Molly, and our family. I have nothing but admiration, and thanks for them for making Molly as comfortable as possible. All the family came home as time came close, and Molly was more or less in a coma, and fading fast. Shortly before the nurse left at around midday, on the 24th July, I was downstairs talking to her, and she told me that Molly had only a short time left. Emotion flooded over me and I had to rush back up to be with Molly.

The nurse then left. Ann and Susan were with Molly, and they noticed that Molly's finger nails were getting black. I had not noticed this, but knew that there was a change for the worse. We were constantly at her side, and Susan was holding Molls hand when she said, "Dad, Mom's nails are going blacker." Susan then left the room, and tried to phone the nurse without success. Immediately, I called everybody to the room, my sisters, and Moll's brother Donald, and her sister Eileen with her husband Michael, who had been downstairs. It was 12.36 p.m. on July 24th, 1995 when Molly drew her last breath, as I held her hand. Ann told me that Molly opened her eyes for a second, just before that, and I wish I had seen her eyes once more. I was bent forward saying a prayer, and I missed that moment. My God! It is so difficult writing, and reliving this part of my life.

I feel resentful to some of the medical profession for what I class, as lack of professional conduct, and attention to detail, and correct analysis of Molly's problem. How could the Hospital in Cork, on their first X-ray find the cancer that was widespread, while the Dublin hospitals had not highlighted any serious problems just prior to the Cork tests? Beaumont did highlight shadows in x-rays which they could not explain and took no further action. Another incident happened when I took Molly to the Mater Public for checks. She felt so weak she could not walk so I sat her down. Various nurses and doctors passed and must have seen her distress and yet did nothing about it. Suddenly, a young nurse came over and asked if she could help. I gratefully asked her to get a wheelchair, which she did, and wheeled Molly to the examination room two floors up. What happened as we entered the room was unbelievable, and made my blood boil. At the desk were three nurses being chatted up by a young doctor. This fellow had all the makings of a, 'Man about town,' trying to impress the girls. He lifted his head when we came into the room, and addressed the kind nurse who had helped. "Why are you up here?" He piped up.

She just said, "I saw this lady was in difficulty and helped her get a chair."

He then put on his bedside manner and said; "You should keep your head down when situations like this face you and pass by. I always do." He bragged in front of Molly and me. I turned on him, felt like punching him, but instead I kept my cool. I said, "You young upstart. You're not fit to be a doctor. You're a disgrace to your profession and need to learn humanity, and keep away from my wife."

He blushed, and went out, whilst the other nurses attended Molly. I again thanked the nurse who helped.

A further incident occurred in this hospital on a second occasion. Again, I brought Molly in after her major surgery, to have a broken staple removed from her neck. The nurse came in, looked, and went off to get a tweezers. She returned after about five minutes to say, she could not find a tweezers, and asked us to come back tomorrow. I lost my temper! As loud as I could, I shouted at her, and asked what kind of hospital this was, that there was not a tweezers available. I then sat down. I told her that I would not move until the broken staple is removed from my wife's neck. Within five minutes it was done, but can you blame me for these outbreaks because of such incompetence? There's more, but I think I have relived enough for now. Molly's sister Anne died shortly after Molly.

With all happiness comes the cost, and it was enormous. After thirty eight years of marriage, and only some months after Susan's wedding, my darling Molly was taken from us, having suffered cancer for a number of years, without being diagnosed in time, and despite after being diagnosed, she underwent radical surgery, was given a clean bill of health, and made countless visits to hospitals. We were all shattered, but have the consolation of knowing she died at home with her loving family, and relatives around her. It took years to get over it.

We are constantly reminded when we see the Red Admiral butterfly. At the gates of the cemetery on July 1995, a Red Admiral butterfly met us, and when we got to the grave it landed on the coffin. Many times over the years, this type of butterfly crosses our paths, and is a source of comfort. I address my butterflies as Molly, when I see them. Quite often, they appear in the church during Mass. Some have come into our bedroom when the window was open. Many of Molly's friends also associate the butterfly with

Molly. She had been such a well-loved and loyal friend. I miss her so much.

May my darling Rest in Peace. Thanks for the memories Molly, darling, and the wonderful family we have.

Oft, in the stilly night,

Ere Slumber's chain has bound me,

Fond Memory brings the light

Of other days around me;

The smiles, the tears,

Of boyhood's years,

The words of love then spoken;

The eyes that shone,

Now dimmed and gone,

The cheerful hearts now broken...

Thomas Moore

1993,
Lynch siblings; Sheila, Patricia, Eileen, Anthony, Jack, Kathleen, Nora, Mary.

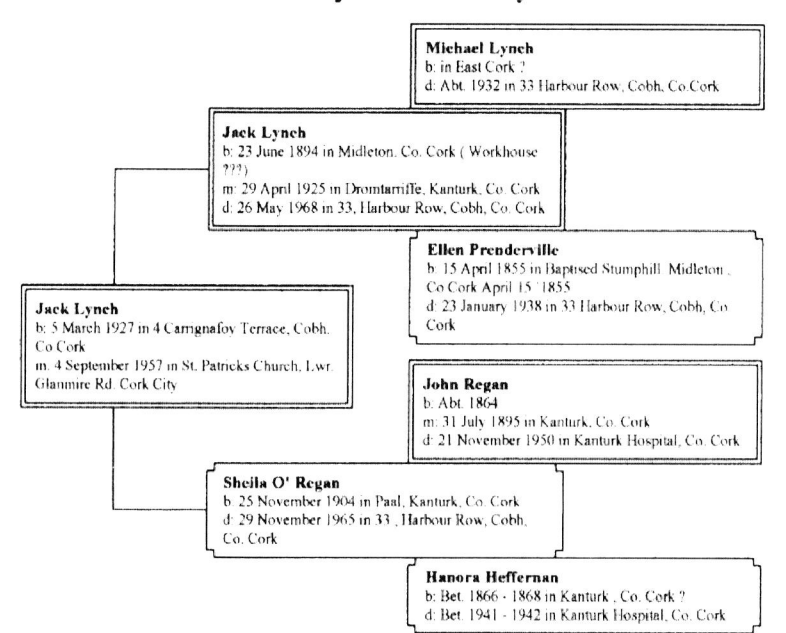

Appendix 1:
Ancestors of John Patrick Lynch

Ancestors of John Patrick Lynch

Michael Lynch
b: in East Cork ?
d: Abt. 1932 in 33 Harbour Row, Cobh, Co.Cork

Jack Lynch
b: 23 June 1894 in Midleton, Co. Cork (Workhouse ???)
m: 29 April 1925 in Dromtarriffe, Kanturk, Co. Cork
d: 26 May 1968 in 33, Harbour Row, Cobh, Co. Cork

Ellen Prenderville
b: 15 April 1855 in Baptised Stumphill, Midleton, Co.Cork April 15 1855
d: 23 January 1938 in 33 Harbour Row, Cobh, Co Cork

Jack Lynch
b: 5 March 1927 in 4 Carrignafoy Terrace, Cobh, Co Cork
m: 4 September 1957 in St. Patricks Church, Lwr. Glanmire Rd. Cork City

John Regan
b: Abt. 1864
m: 31 July 1895 in Kanturk, Co. Cork
d: 21 November 1950 in Kanturk Hospital, Co. Cork

Sheila O' Regan
b: 25 November 1904 in Paal, Kanturk, Co. Cork
d: 29 November 1965 in 33, Harbour Row, Cobh, Co. Cork

Hanora Heffernan
b: Bet. 1866 - 1868 in Kanturk, Co. Cork ?
d: Bet. 1941 - 1942 in Kanturk Hospital, Co. Cork

Appendix 2:
Descendents of John Patrick Lynch

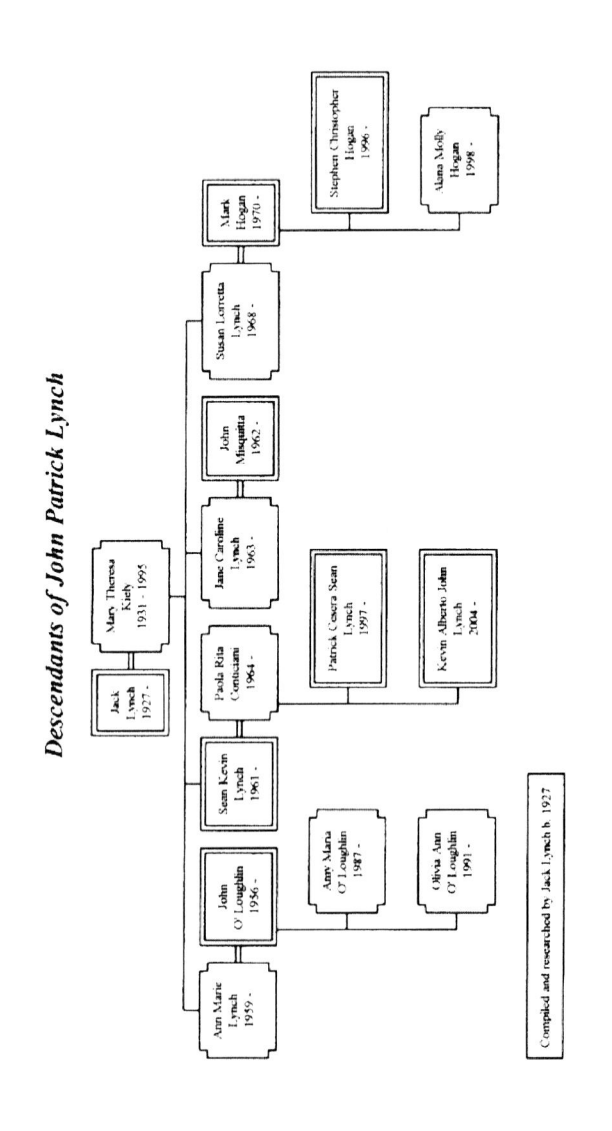

Descendants of John Patrick Lynch

- Jack Lynch 1927 - m. Mary Theresa Kiely 1931 - 1995
  - John O'Loughlin 1956 - m. Ann Marie Lynch 1959 -
    - Ann Maria O'Loughlin 1987 -
    - Olivia Ann O'Loughlin 1991 -
  - Sean Kevin Lynch 1961 -
  - Paola Rita Contessani 1964 -
    - Patrick Cesera Sean Lynch 1997 -
    - Kevin Alberto John Lynch 2004 -
  - Jane Caroline Lynch 1963 -
  - John Misquitta 1962 -
  - Susan Loretta Lynch 1968 - m. Mark Hogan 1970 -
    - Stephen Christopher Hogan 1996 -
    - Alana Molly Hogan 1998 -

Compiled and researched by Jack Lynch b. 1927

Appendix 3:
History of Ships and Ports I signed on and off.

Name of ship	Net tons	Type of ship	I signed on.	I signed off.
Isle of Guernsey	2152.3	Passenger Ferry	Southampton, Hampshire. 08/07/1949	Southampton. Hampshire. 13/10/1949
Winkleigh	8909	Tramp steamer	Hull, Yorkshire. 17/10/1949	Popular, London. 05/01/1950
Tower Hill	5067	Tramp steamer	Tilbury, London 23/01/1950	Birkenhead, Lancs. 19/07/1950.
Lampania	3625	Shell tanker	Ellesmere Port, Lancs. 03/08/1950	Birkenhead, Lancs. 11/10/1950.
President Brand	4751	T2 tanker	Ellesmere Port, Lancs. 27/10/1950	Heysham, Lancs. 12/12/1950.
Orford	4210	Liberty Park tramp.	Manchester, Lancs. 02/01/1951.	Falmouth, Cornwall. 23/01/1952
Esso Birmingham	6324	Esso tanker	Southampton, Hampshire. 14/03/1952	Bremerhaven, Germany. 21/08/1952.
Sedgepool	4459	Liberty tramp	West Hartlepool, Durham 15/09/1952	North Shields, Tyne 31/08/1953
Goulistan	5058.3	Passenger & cargo	Liverpool, Lancs. 05/10/1953	Liverpool, Lancs. 11/03/1954
Selector	4743	Tramp steamer	Liverpool, Lancs. 13/03/1954	Liverpool, Lancs. 07/07/1954
Albert G. Brown	4475	Liberty tanker	Manchester, Lancs. 19/03/1955.	Manchester, Lancs. 17/02/1956.
Albert G. Brown	4475	Liberty tanker	Dover, Kent. 17/07/1956.	San Pedro, Calif. 12/06/1957.
Albert G. Brown	4475	Liberty tanker	Rotterdam, Holland. 16/10/1957...	Houston, Texas. 22/01/1958.
Adellen	5786	Tanker	South Shields, Tyne. 13/06/1958	Jacksonville, Fla. 10/12/1958.
Turmoil	1136.2	Deep sea Tug	Cobh, Co. Cork. Ire. 06/05/1959	Milford Haven, Lancs. 18/06/1959.
Kiaora	4751	T2 Tanker	Killingholme, Scot. 18/07/1959	Grangemouth, Scot. 10/11/1959

Appendix 4:
History of my voyages

PORT	COUNTRY	CONTINENT	SHIPS' NAMES
Accra	Ghana	Africa	Winkleigh.
Aden	Yemen	Asia	Tower Hill, Goulistan.
Amsterdam	Holland	Europe	Winkleigh, Esso Birmingham, Orford.
Antwerp	Belgium	Europe	Lampania, Orford.
Aruba	Neth. Antilles	Caribbean Sea	Albert G. Brown, Adellen.
Bahia Blanca	Argentina	South America	Orford.
Bahrein	Bahrein	Persian Gulf, Asia	President Brand, Goulistan.
Balboa	Panama	Central America	Orford.
Bari	Italy	Europe	Tower Hill.
Basterre	British Honduras	Central America	Selector.
Basra	Iraq	Persian Gulf, Asia	President Brand.
Belize	British Honduras	Central America	Selector.
Birkenhead	UK	Europe	Tower Hill, Lampania.
Bitter Lakes	Egypt	Africa	Tower Hill, Goulistan, President. Brand, Esso Birmingham.
Bordeaux	France	Europe	Orford.
Bremen	Germany	Europe	Lampania.
Bremerhaven	Germany	Europe	Esso Birmingham, Brest.
Brest	France	Europe	Orford.
Buenos Aires	Argentina	South America	Orford.
Canary Islands	Spain	Atlantic	Winkleigh.
Caracas	Venezuela	South America	Albert G. Brown, Adellen.
Casablanca	Spanish Morocco	North Africa	Goulistan.
Ceuta	Spanish Morocco	North Africa	Goulistan.
Chesapeake Bay, Del.	USA	North America	Albert G. Brown.
Clyde	Scotland	Europe	Kiora.
Cobh	Ireland	Europe	Turmoil
Colon	Panama	Central America	Albert G. Brown, Sedgepool.
Corpus Christi, Texas	Texas, USA	North America	Albert G. Brown, Sedgepool.
Cristobal	Panama	Central America	Albert G. Brown, Sedgepool.
Cul de Sac Bay	St. Lucia	Caribbean Sea	Selector
Curacao	Neth. Antilles	Caribbean Sea	Albert G. Brown, Lampania.
Delaware Bay	USA	North America	Albert G. Brown.
Dominica	Windward Islands	Caribbean Sea	Selector.
Dover	UK	Europe	Albert G. Brown, Kiora.
Ellesmere Port	UK	Europe	Albert G. Brown, President Brand, Lampania.
Falmouth	UK	Europe	Orford, Esso Birmingham.

360

Fort Lauderdale, Fla	Florida, USA	North America	Albert G. Brown, Adellen.
Freemantle	Australia	Australia	Tower Hill.
Freetown	Sierra Leone	West Africa	Winkleigh.
Galveston	Texas	North America	Albert G. Brown, Sedgepool.
Geelong	Australia	Australia	Tower Hill.
Genoa	Italy	Europe	Sedgepool.
Gibraltar	Britain	Mediterranean Sea	Turmoil.
Gotenburg	Sweden	Europe	Lampania.
Grangemouth	Scotland	Europe	Kiaora.
Grimsby	UK	Europe	Kiaora.
Guaymas	Mexico	North America	Sedgepool.
Guernsey	Britain	Europe	Isle of Guernsey
Hamburg	Germany	Europe	Esso Birmingham, Orford, Albert G. Brown
Honolulu	Hawaii,USA	Pacific Ocean	Sedgepool.
Heysham	UK	Europe	President Brand
Houston, Texas	USA	North America	Albert. G. Brown, Selector, Adellen.
Hull	UK	Europe	Winkleigh, Selector, Goulistan.
Imjuiden	Holland	Europe	Esso Birmingham, Winkleigh.
Jacksonville, Florida	USA	North America	Albert G. Brown, Selector.
Kamashi	Japan	Far East, Asia	Sedgepool.
Kingstown	St. Vincent Island	Caribbean Sea	Selector.
Kuwait	Kuwait	Asia (Persian Gulf)	Esso Birmingham.
Lagos	Nigeria	West Africa	Winkleigh.
Las Palmas	Canary Is. Spain	Atlantic Ocean	Winkleigh
River Plate	Argentina/Uruguay	South America	Orford.
La Spezia	Italy	Europe	Turmoil.
Lima	Peru	South America	Sedgepool.
Liverpool	UK	Europe	Orford, Goulistan, Selector, Albert G. Brown.
London	UK	Europe	Winkleigh, Tower Hill, Esso Birmingham.
Long Beach, California.	USA	North America	Albert G. Brown.
Lynchburg, Virginia	USA	North America	Albert G. Brown.
Manchester	UK	Europe	Albert G. Brown, Selector.
Maracaibo Bay	Venezuela	South America	Albert G. Brown, Esso Birmingham.
Martinique	Winward Islands	Caribbean Sea	Selector.
Mazatlan	Mexico	North America	Sedgepool.
Melbourne	Australia	Australia	Tower Hill.
Miami, Florida.	USA	North America.	Sedgepool.
Milford Haven	UK	Europe	Turmoil.

Monserrat	Leeward Islands	Caribbean Sea	Selector.
Naples	Italy	Europe	Tower Hill.
Newark, New Jersey	USA	North America	Sedgepool.
New Orleans, Louis'ana	USA	North America	Albert G. Brown.
Newport, Mon.	Wales, UK	Europe	Albert G. Brown, Adellen.
Newport News	USA	North America	Albert G. Brown.
New Vitas	Cuba	Caribbean Sea	Albert G. Brown, Adellen.
New York, NY	USA	North America	Sedgepool.
Norfolk, Virginia.	USA	North America	Albert G. Brown.
North Shields	UK	Europe	Sedgepool.
Oakland, California	USA	North America	Sedgepool.
Oran	Algeria	North Africa	Goulistan.
Perth	Australia	Australia	Tower Hill.
Philadelphia, Pn	USA	North America	Sedgepool.
Port Everglades,	USA	North America	Albert G. Brown.
Portland, Oregon.	USA	North America	Albert G. Brown.
Port Louis	Mauritius	Indian Ocean	Goulistan.
Port Said	Egypt	Africa	Tower Hill, Goulistan, Esso Birmingham, President Brand
Portsmouth, Nh.	USA	North America	Albert G. Brown
Prince Rupert Is.	Canada	North America	Sedgepool.
Puerto Rico	USA Protectorate	Caribbean Sea	Albert G. Brown.
Ras Tanura	Saudi Arabia	Asia, Persian Gulf	Esso Birmingham, President Brand.
Richmond, Virginia.	USA	North America	Albert G. Brown.
Rotterdam	Holland	Europe	Esso Birmingham, Lampania, Albert G. Brown.
Rouen	France	Europe	Albert G. Brown, Orford.
St. Helier, Jersey.	Channel Islands, U	Europe	Isle of Guernsey.
St. John,	Antigua Island	Caribbean Sea	Selector.
St. Kitts	Leeward Islands	Caribbean Islands	Selector.
St. Lucia	Windward Islands	Caribbean Sea	Selector.
St. Peter Port, Guernsey	Channel Islands, U	Europe	Isle of Guernsey.
St. Vincent	Winward Islands	Caribbean Sea	Selector
San Diego, California.	USA	North America	Albert G. Brown.
San Francisco, Calif.	USA	North America	Albert G. Brown.
San Juan	Peru	South America	Sedgepool.
San Pedro, California.	USA	North America	Albert G. Brown.
Savannah, Georgia	USA	North America	Albert G. Brown.
Seattle, Washington.	USA	North America	Albert G. Brown.
Southampton	UK	Europe	Isle of Guernsey, Esso Birmingham.

South Shields	UK	Europe	Adellen.
Stockholm	Sweden	Europe	Esso Birmingham.
Suez Canal	Egypt	Africa	Tower Hill, Goulistan, President Brand, Esso Birmingham.
Takoradi	Ghana	West Africa	Winkleigh.
Tampa, Florida	USA	North America.	Albert G. Brown.
Thane	India	Asia	Esso Birmingham.
Trenton, New Jersey	USA	North America.	Sedgepool.
Tilbury, London	UK	Europe	Winkleigh, Tower Hill, Esso Birmingham.
Uum Said	Qatar	Persian Gulf	Goulistan.
Vancouver	Canada	North America	Sedgepool.
Vita	Cuba	Caribbean Sea	Albert G. Brown.
Warri	Nigeria	West Africa	Winkleigh.
West Hartlepool	UK	Europe	Sedgepool
Wilmington, Delaware.	USA	North America	Albert G. Brown.
Winneba	Ghana	West Africa	Winkleigh.

Appendix 5:
Radio and seagoing documents

My duplicate Radio Officer's Certificate

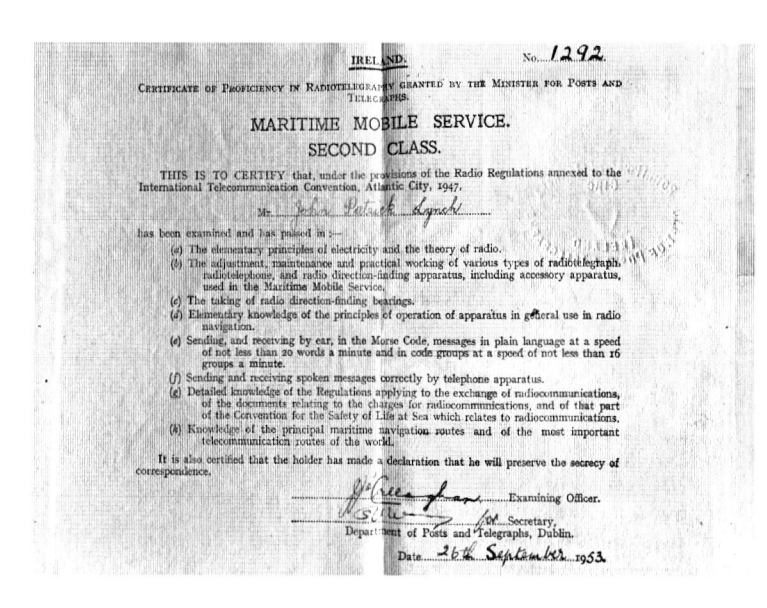

My Postmaster General Certificate

My Liberian Radio Certificate

My British seaman's record book

Records of dates, British ships, ports, where I signed-on, and signed off.

Records of dates, British and South African flag ships, ports, where I signed-on, and signed off.

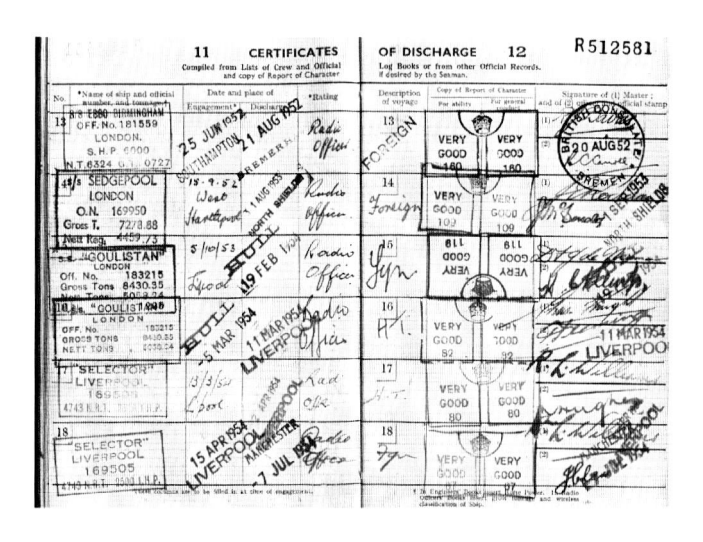

Records of dates, British flag ships, ports, where I signed-on, and signed off.

Records of dates, British flag ships, ports, where I signed-on, and signed off.

Records of dates, Liberian flag ships, ports, where I signed-on, and signed off.

CERTIFICATE OF DISCHARGE

S/S S.S. ALBERT G. BROWN of MONROVIA, LIBERIA
Name of Seaman JOHN LYNCH Rating R/O
Date and Place of Shipment TAVER ENG JULY 17TH 1956
Date and Place of Discharge SAN PEDRO CAL JUNE 12TH 1957
Gross Tons 7289
Nature of Voyage(s) FOREIGN Net Tons 4475
Type of Vessel LIBERTY STEAM TANKER Official Number 406
Signature of Seaman John P Lynch
Signature of Master Julius B. Tico

CERTIFICATE OF DISCHARGE

S/S ALBERT G. BROWN of MONROVIA, LIBERIA
Name of Seaman JOHN P. LYNCH Rating RADIO OFFICER
Date and Place of Shipment OCTOBER 14th, 1957 ROTTERDAM
Date and Place of Discharge JANUARY 22nd, 1958 HOUSTON, TEXAS
Gross Tons 7289
Nature of Voyage(s) FOREIGN Net Tons 4475
Type of Vessel STEAM TANKER Official Number 406
Signature of Seaman
Signature of Master

Records of dates, Liberian flag ships, ports, where I signed-on, and signed off.

CERTIFICATE OF DISCHARGE

S/S ADELLEN of MONROVIA, LIBERIA
Name of Seaman: JOHN P. LYNCH   Rating: RADIO OFF./ PURSER
Date and Place of Shipment: SOUTH SHIELDS, ENG. JUNE 13th. 1958
Date and Place of Discharge: JACKSONVILLE, FLA. NOV. 22nd. 1958
Gross Tons: 9501
Nature of Voyage(s): FOREIGN   Net Tons: 5936
Type of Vessel: STEAM TANKER   Official Number: 1266
Signature of Seaman:
Signature of Master:

CERTIFICATE OF DISCHARGE

S/S KIAORA of MONROVIA, LIBERIA
Name of Seaman: JOHN P. LYNCH   Rating: RADIO OFFICER/PURSER
Date and Place of Shipment: JULY 18th. 1959 KILLINGHOLME, ENGLAND.
Date and Place of Discharge: NOVEMBER 10th. 1959 GRANGEMOUTH, SCOTLAND.
Gross Tons: 7972
Nature of Voyage(s): FOREIGN   Net Tons: 4751
Type of Vessel: STEAM TANKER   Official Number:
Signature of Seaman:
Signature of Master:

Records of dates, Liberian flag ships, ports, where I signed-on, and signed off.

USA Immigration landing permit in US